THE GREATEST SOURCE OF ENERGY

A New Theory of Time

THE GREATEST SOURCE OF ENERGY

A New Theory of Time

Lamont Williams

McNair & Williams Publishing, Baltimore, Maryland

The Greatest Source of Energy—A New Theory of Time

McNair & Williams Publishing
Baltimore, Maryland

Send correspondence to:
publisher@mcnairandwilliams.com

Printed in the United States of America

ISBN: 9781982006600 (paperback)

ISBN: 9780984450374 (hardcover)

Library of Congress Control Number: 2010912246

This book is dedicated to children everywhere.
May we all give them the peaceful world they deserve.

It is also dedicated to my mother, Brenda; my father, Clyde;
my brothers, André and Shawn; and the rest of my family.
Thank you for your love and support.

The book is further dedicated to educators worldwide.
You help to uplift humanity.

DISCLAIMER

The work presented here provides a method of combining General Relativity, Quantum Mechanics, and other scientific concepts. Although it is described as a "theory" for practical purposes, its mathematical formulation has not yet been fully established. As such, it is likely more accurately described as an "idea for a theory" rather than a theory itself. The method also has not been tested as of its publication date and, therefore, no claim can be made that it is an accurate description of time, energy, or any other aspect of the physical world. Additionally, every effort has been made to ensure accuracy throughout the text. However, mistakes may exist, and they are the sole responsibility of the author, not the reviewers or editors. This text should be used only as a general introduction to the new theory and not as an ultimate source for the other scientific concepts discussed. Also, the book contains information that is current only as of the printing date.

PREFACE

We live in exciting scientific times. Technological advances today are allowing us to dig deeper into the universe than ever before and are simultaneously allowing us to extend our senses ever further into the reaches of space. Scientific progress marches forward, and this progress is indeed the heart of human advancement. Our increasing scientific knowledge continuously enhances such areas as art, athletics, business, and the humanities. Yet, despite the vast knowledge we have, there is still much to learn. It is my strong hope that the book presented here will shed some new light on many long-lasting questions, problems, and mysteries in the scientific arena and help us advance yet another step in our scientific understanding.

— Lamont Williams

CONTENTS

INTRODUCTION

Time is the most mysterious thing in the universe. For thousands of years, it has lain outside the grasp of human understanding. The most we have been able to say about time in everyday language is that it passes—sometimes quickly, sometimes slowly—and that it is different from space. We know it exists, but we have not been able to say definitively what it is or how it works. In fact, the two most successful theories describing the universe—General Relativity and Quantum Mechanics—have completely different views on the subject. In General Relativity, time is considered to be a dimension that is "melded" to the three dimensions of space. Together, space and time are considered to constitute a four-dimensional field (three spatial dimensions plus one time dimension) called spacetime. According to the theory, space-time is an active player in the universe. It influences and responds to physical events: Two often-quoted statements by noted scientist John Wheeler are that "spacetime tells matter how to move" and "matter tells spacetime how to curve."[1] In Quantum Mechanics, space and time are considered to be separate things and to have no involvement in physical events. Space is considered to be a flat, fixed arena in which events just happen, and time a mysterious, separate thing that ticks away regularly "somewhere out there."

For many years, scientists have sought to combine General Relativity and Quantum Mechanics into a single theory. This is because, to date, all of the observed fundamental phenomena in the universe are described by one or the other of these two theories, so by combining them, they would have one theory describing everything—all observed fundamental phenomena. Therefore, they call this single theory the Theory of Everything. Despite a great deal of work in this area, scientists have not been able to develop this theory. As might be expected, the principal problem that emerges concerns time: When scientists combine the mathematics of General Relativity with the mathematics of Quantum Mechanics, time vanishes. That is, they get an equation called the Wheeler-DeWitt equation—named after John Wheeler and Bryce DeWitt, another noted scientist—that basically says time does not exist in the universe. They call this problem, naturally, the problem of time. However, not only do we intuitively understand that time exists, time plays an important role in both General Relativity and Quantum Mechanics, so an equation that is born from the combination of these two theories that says the universe has no time is puzzling. What's wrong here?

The answer is nothing. Time should vanish from the math when General Relativity and Quantum Mechanics are combined. The Wheeler-DeWitt equation, with its "time equals zero" solution, indeed provides the clearest and most complete picture of time in the universe. The reason many people have found this puzzling is because they have not understood one very important thing, that General Relativity and Quantum Mechanics are actually describing the same process—that is, the same *temporal* process—but from opposite perspectives. Therefore, when the two theories are combined, the math, through the Wheeler-DeWitt equation, appears to indicate that there is zero time in the universe. Consider this: If I were to walk east five steps, turn around, walk west five steps, and then turn back around, mathematically it would look like I traveled zero steps; after all, at the end of the process, I would still be at my original starting point. However, I would have actually traveled more than ten steps. This is basically the situation with regard to General Relativity, Quantum Mechanics, and the Wheeler-DeWitt equation concerning time. It is not that time does not exist; there are simply two equal, yet opposite temporal processes occurring simultaneously in the universe. General Relativity describes one; Quantum Mechanics describes the other. Put them together, and they cancel each other out mathematically. Not understanding that General Relativity and Quantum Mechanics are opposite sides of the same temporal coin has been the principal obstacle preventing the successful combination of these two theories into the Theory of Everything.

In this book, a new theory is described—the *Temporal Energy Theory*. Without going into mathematical detail, it presents a new theory of time and, through that theory, adds to the prevailing interpretations of General Relativity and Quantum Mechanics, showing how these two theories might be successfully combined. The book also uses the theory to connect the dots between additional scientific concepts and, in the process, build a consistent model of how the universe likely operates on the small and grand scales.

PART
1

TOWARD A THEORY
OF EVERYTHING

CHAPTER
1

INTRODUCTION TO THE
TEMPORAL ENERGY THEORY

I n Temporal Energy Theory (TET), time is considered to be associated with extremely small particles, called *temporal particles*, that are emitted from space and move faster than the speed of light. Indeed, space itself moves faster than the speed of light. In this theory, the universe is analogous to an electrical circuit involving a battery, where space is akin to the battery, matter is akin to the device on which the battery is operating, and time is akin to the electrical current *(Figure 1.1)*.

A temporal particle (designated as *t*) is considered to have one of three spatiotemporal charges (+, 0, −). The spatiotemporal charge relates to how a temporal particle will interact with space and matter. Space emits t^+, which travels toward matter. Through contact with matter, t^+ is converted first to t^0

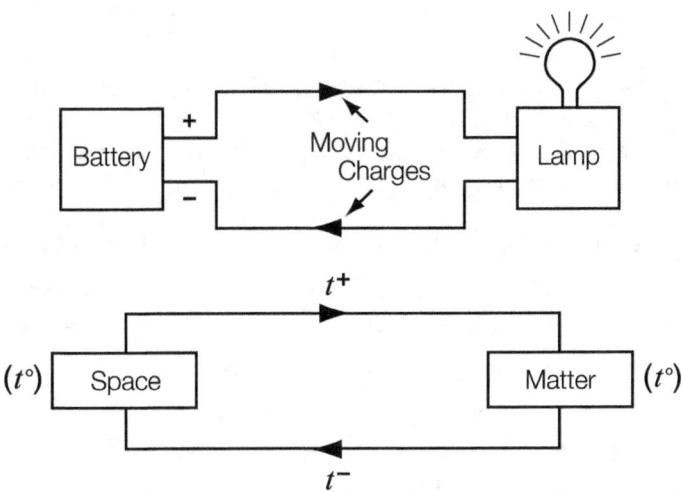

Figure 1.1 *In Temporal Energy Theory (TET), the universe is analogous to an electrical circuit involving a battery, where space is akin to the battery, matter is akin to the device on which the battery is operating, and time is akin to the electrical current.*

and then to t⁻, which disassociates from matter and is absorbed by space. In space, t⁻ is converted first to t^0 and then back to t⁺, which is then reemitted from space. The process continually repeats. The conversion of t⁺ to t⁻ (by matter) and t⁻ to t⁺ (by space) is called *temporal respiration* or the *temporal-conversion process*, and represents the passage of time. From the perspective of matter, t⁺ symbolizes the future, t^0 the present, and t⁻ the past. From the perspective of space, it is the other way around: t⁻ symbolizes the future, t^0 the present, and t⁺ the past. This is the origin of the two opposing temporal processes that were mentioned in the Introduction and that relate to General Relativity and Quantum Mechanics. In the simplest of senses, General Relativity focuses on the t⁺ to t⁻ process, whereas Quantum Mechanics focuses on the t⁻ to t⁺ process. To make things easier in this book, the t⁺ to t⁻ process will be written t⁺/t⁻, and the t⁻ to t⁺ process will be written t⁻/t⁺. Already, using these TET concepts, it can be understood how an answer of "zero" would result from combining the two processes *(Figure 1.2)*.

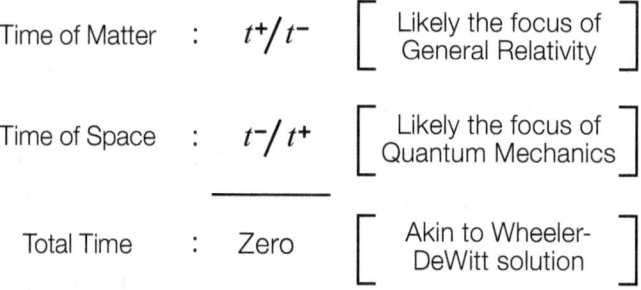

Time of Matter	:	t^+/t^-	Likely the focus of General Relativity
Time of Space	:	t^-/t^+	Likely the focus of Quantum Mechanics
Total Time	:	Zero	Akin to Wheeler-DeWitt solution

Figure 1.2 *From the TET perspective, General Relativity focuses on the t^+/t^- temporal-conversion process, whereas Quantum Mechanics focuses on the t^-/t^+ temporal-conversion process. These two processes cancel each other mathematically, leaving a solution of zero and making it seem as though there is no time in the universe.*

In the chapters that follow, TET is discussed in relation to General Relativity and Quantum Mechanics. And it is shown how TET can be used to unite these two theories. More precisely, it is shown how TET can be used to describe all of the fundamental forces of the universe known to date (forces described well by General Relativity or Quantum Mechanics—but not both). These forces are gravity (described by General Relativity) and the electromagnetic, strong nuclear, and weak nuclear forces (described by Quantum Mechanics). The various roles TET plays in other major phenomena in the universe, for example mass, electric charge, atoms, and the Big Bang, are also discussed.

CHAPTER
2

GENERAL RELATIVITY

G eneral Relativity (GR) is the theory created by Albert Einstein that describes the force of gravity—the attraction massive objects have for each other, the force that keeps our feet firmly on the ground. According to the theory, space and time are treated as a single, four-dimensional field called spacetime—three dimensions of space plus one dimension of time. Although not palpable (capable of being felt), spacetime is considered to be malleable (capable of being stretched, compressed, bent, or "curved"): A massive object such as the sun is said to curve the spacetime around it. And these curves affect the movement of other objects in its vicinity. To reiterate the statements from John Wheeler, "spacetime tells matter how to move," and "matter tells spacetime how to curve."[1] Objects in the vicinity of a more massive object are thought to move along the curves it produces in spacetime. Thus, the Earth, in orbiting the sun, would actually be following the curves in spacetime that the sun creates.

Often, in attempting to demonstrate gravity from the perspective of GR, individuals place a ball onto a flat, flexible sheet. The sheet represents spacetime, and the ball represents the sun. Next, they place a smaller, second ball next to the previous one—the smaller ball represents the Earth. The second ball rolls, or falls, into the first because the first ball has created a curve in the flexible sheet, that is, in the spacetime fabric. If the smaller ball were constantly revolving around the larger one, the way the Earth orbits the sun, the smaller ball would not completely fall into the larger ball, but it would still follow the path of the curve in the sheet created by the larger ball (*Figure 2.1*). This, many people say, is how gravity generally works according to GR. They are attempting to demonstrate the GR-related idea that it is the curvature of spacetime that brings about the phenomenon we call gravity.

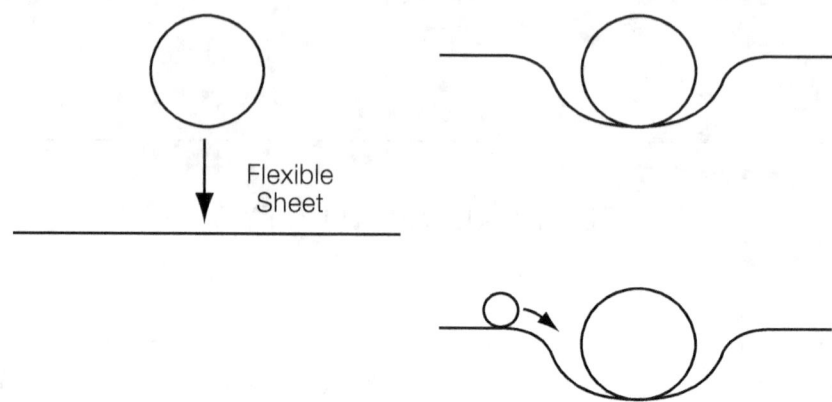

Figure 2.1 *To demonstrate the General Relativity (GR) idea that gravity is caused by the curvature of spacetime, a large ball is placed onto a flexible sheet, then a second, smaller ball is placed there. The smaller ball rolls or "falls" into the larger ball due to the curvature in the sheet/the spacetime created by the larger one.*

Although useful to some degree, this demonstration has unfortunately given many people the impression that GR says that gravity is caused by the curvature of the spatial aspect of spacetime, when actually the theory says that gravity is caused by the curvature of the temporal aspect of spacetime, not the spatial. Too often in discussions of GR, the concept of spacetime is taken so lightly that individuals begin to think of space and spacetime as the same thing, when in fact they are very different; as a consequence, many people begin to lose sight of time and the role it plays. The idea that gravity is caused by the curvature of the time dimension of spacetime was emphasized by Einstein in his book, *The Meaning of Relativity*.[2] Einstein and other experts in GR have stressed that while all four dimensions of spacetime are indeed curved by matter, what causes an object to fall for example toward Earth are curves in the time dimension, not the spatial dimensions.[3,4,5,6] It is just that in GR, it is almost meaningless to talk about time and space separately. References to spacetime as a single thing unto itself are typically seen, rather than references to space apart from time or time apart from space. If time is curved, space is curved; if space is curved, time is curved. Space and time in GR are considered to be components of a single unified field, but spatial curves do not cause gravity in GR.

Gravity and Temporal Energy Theory

Unlike GR, which considers gravity to be caused by the curvature of the so-called time dimension of spacetime, TET considers gravity to be caused by the flow of temporal particles toward matter and ultimately back into space. This of course involves the temporal respiration of matter (the conversion of t^+ to t^-) and the subsequent absorption of t^- by space. The volume of space in which we live should be thought of as the storage place for t^+ particles. To help with this, you should think of there being two types of space: interior and exterior, with interior space being the space in which we live and the storage place for t^+ particles and exterior space being the storage place for t^- particles. (Note that there is no life in exterior space. These two types of space will be further discussed in subsequent chapters.) With the addition of these concepts, the cycle introduced in chapter 1 can be revised (*Figure 2.2*). It is in exterior space that t^- particles are converted to t^+ particles by *exterior space* itself. They are subsequently ejected into interior space where they are converted by *matter* back into t^- particles and are then reabsorbed by exterior space. Interior space is called what it is only because we, living within that space, see the inside or "interior" of that space. The other space is outside of our direct experience—it is exterior to us—and thus is termed exterior space. Note, however, that there is nothing really "interior" about interior space or "exterior" about exterior space. The spaces are essentially superimposed on each other, parallel.

Positive temporal particles not only fill all of interior space, they also form bonds with each other and thus form a single, unified field throughout that

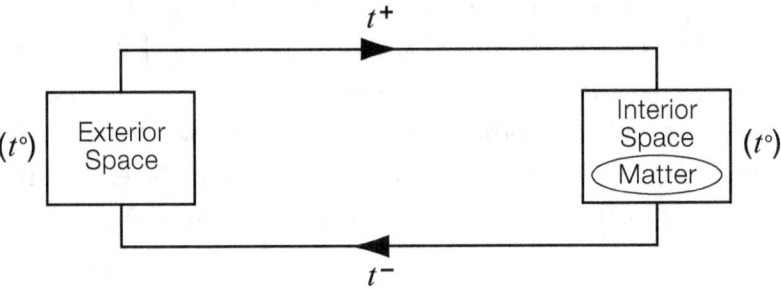

Figure 2.2 *A revision of the schematic in Figure 1.1. A t^+ particle emerges from exterior space and enters interior space where it interacts with matter. Through its contact with matter, the particle is converted ultimately to t^- and is reabsorbed by exterior space.*

space. They are also in constant motion, moving faster than the speed of light. This is analogous to water in that individual water molecules are constantly in random motion but also establish bonds with other water molecules, forming a well-defined liquid. Much of t^+ motion is also random, but there is a non-random aspect to this motion, as well. As noted above, t^+ particles move toward matter. To be more precise, they are attracted to and hence move toward all forms of energy. Positive temporal particles swarm around a matter particle, like moths swarm around a flame, because it has energy. If the matter particle is stationary, this energy is mostly its mass. (Note that this chapter deals solely with matter particles. Gravitational attraction involving other energetic species is discussed in chapter 5.)

The concentrated field of t^+ particles surrounding a matter particle is its gravitational field. However, because all of the t^+ particles in interior space are connected, the matter particle's gravitational field can actually be said to extend throughout all of interior space. To distinguish, let us call the concentrated field of t^+ particles immediately surrounding a matter particle its *gravitational field*, and the whole field extending throughout all of interior space, the *gravitational field proper*. Thus, for all matter particles everywhere in interior space, there is only one gravitational field proper. No one particle can claim true ownership.

Note that the density of temporal particles composing the gravitational field of a matter particle increases toward the matter particle's surface, because temporal particles try to get as close to a source of energy (for example, the matter particle's mass) as possible, giving the gravitational field a layered look (*Figure 2.3*). Mass plays an important part in gravity, in that it is on a matter particle's mass that the gravitational field anchors itself. That is, in addition to forming bonds with each other, temporal particles form bonds with a matter particle's mass. Note that the more compact temporal particles are, the stronger the bonds between them and also the mass of the matter particle they are connected to.

Soon after coming into contact with the matter particle, some t^+ particles are transformed into t^-, through the t^+/t^- temporal-respiration process. As the t^-

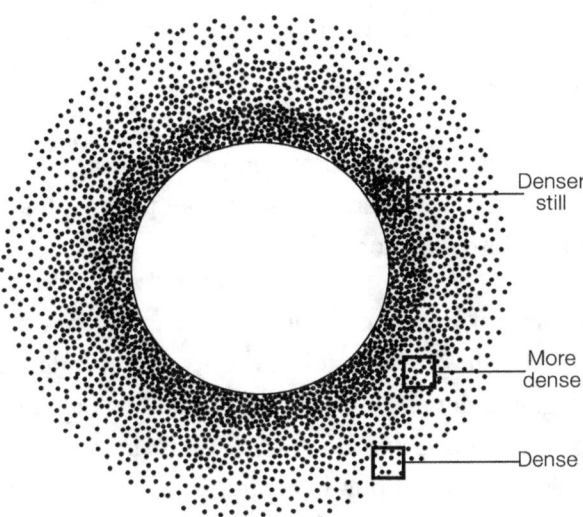

Denser
still

More
dense

Dense

Figure 2.3 *Temporal particles try to get as close to a source of energy as possible, with their density steadily increasing toward the surface of the object.*

particles are absorbed by exterior space, t^+ particles that were further out in the gravitational field move in closer toward the matter particle, filling the open spaces. If the incoming t^+ particles are also linked to a common gravitational field that exists between the previously mentioned matter particle and another matter particle, a tension will develop within that field, as all of the events described above are occurring on that other particle of matter, as well. As they "consume" the t^+ particles closest to them, each matter particle will slowly move toward the other—toward the center of the common field. This is essentially how gravity operates according to TET. (The term *common gravitational field* is used loosely here because the gravitational field proper is the ultimate common field between all matter particles.) The following examples illustrate these ideas further, using larger objects composed of matter particles.

First, to summarize the points above, gravity is caused by these factors in TET:

1. The attraction of t^+ particles to particles of matter (or energy generally);

2. The bonds t^+ particles form with each other and with a matter particle's mass—the more compact the temporal particles are, the stronger the bonds between them and also the mass of the matter particle they are connected to;

3. The conversion of t^+ to t^- by matter;

4. The absorption of t^- by exterior space.

All of these factors concern the flowing of temporal particles toward matter and ultimately their disappearance into the fabric of space itself—out of interior space and into exterior space.

Consider two celestial bodies of about equal mass. They are each surrounded by a concentrated field of t^+ particles, their gravitational fields. The temporal particles in the gravitational fields of the two bodies, like all temporal particles, commingle and form bonds with each other, forming a common gravitational field between them. As temporal respiration occurs in the two bodies, this common field, which is anchored on the mass of the two bodies, becomes stretched as the temporal particles composing it are consumed by the bodies. That is, the bonds between the temporal particles in the shared field begin to become strained as the t^+ particles closest to the two bodies become converted to t^- and are absorbed by exterior space. The field begins to appear like a rubber sheet, the two ends of which are being pulled in opposite directions. The

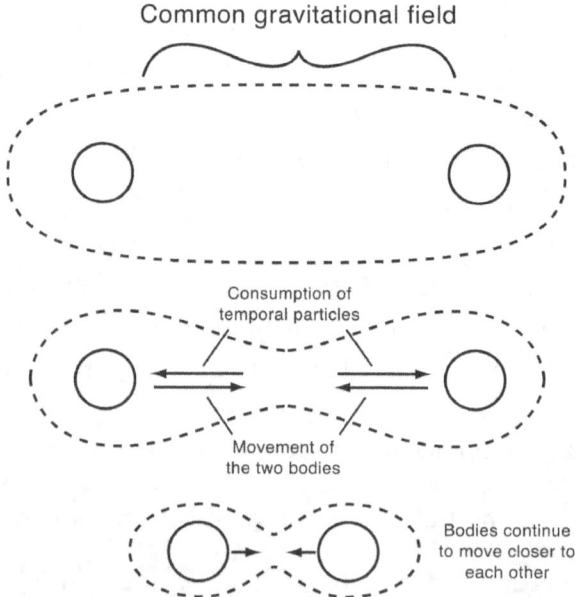

Common gravitational field

Consumption of
temporal particles

Movement of
the two bodies

Bodies continue
to move closer to
each other

Figure 2.4 *In TET, gravity occurs through the consumption of temporal particles in the common gravitational field between two bodies.*

A has a greater pull on B
than B has on A

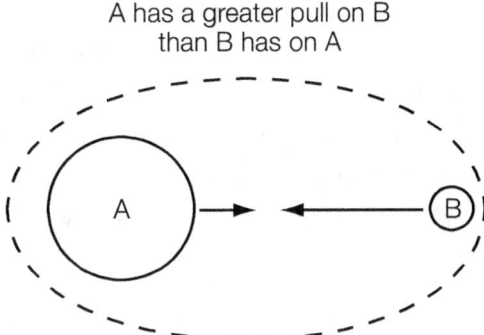

Figure 2.5 *A graphical representation of gravitational attraction between two bodies of unequal mass from the TET perspective.*

temporal-respiration process, absorption of t^- particles, and tension in the shared gravitational field between the bodies cause the two objects to move increasingly closer to each other, until eventually all of the temporal particles in their shared gravitational field have been consumed and they are touching (*Figure 2.4*).

Figure 2.5 shows the gravitational attraction between two bodies of unequal mass. In this example, the more massive body, A, has a greater pull than the less massive one, B, because body A has a larger gravitational field. Let us call body A the sun, and body B the Earth. The reason the Earth does not fall into the sun is because it revolves around the star. The orbital motion of the Earth prevents a full gravitational attraction between the two bodies. In TET, this orbital motion, as well as the rotation of the Earth and sun about their axes, disrupts the bonds between some of the temporal particles in the common field. As stated earlier, temporal particles are attracted to all forms of energy. When the Earth revolves around the sun and when the Earth and sun rotate about their axes, temporal particles move with them due to this motion energy. In some cases, these particles move in opposite or near-opposite directions away from each other, breaking or weakening the bonds between them, which in turn weakens the gravitational attraction between the two celestial bodies. With this weakened attraction and the right distance, direction, and speed, Earth can, and did of course, develop a stable orbit around the sun. Indeed, any massive body can develop such an orbit around another more massive one floating in space, under the right conditions.

Stable orbits are advantageous for obvious reasons, but falling is also an important phenomenon. And it is worth exploring through TET concepts not only why objects fall, for example to the surface of the Earth, but why relatively small objects fall toward the planet at the same rate regardless of differences in their mass. For instance, ignoring such factors as air resistance, a bowling ball and a feather if dropped from the same height at the same time will reach the ground at exactly the same moment.

First, falling and gravitational attraction are the same thing when no other force is playing a role in the falling action. Thus, in the first example (*Figure 2.4*), the celestial bodies were falling into each other, and this would happen with regard to the Earth and sun also, if again the Earth did not revolve around the star. Therefore, an object falls into another object for the same reasons that gravity occurs in general. Second, the reason relatively small objects will fall into vastly larger ones at the same rate, from the TET perspective, is that the more massive a falling object is, the more temporal particles there are mediating the attraction between it and the object it is falling into. The less massive a falling object is, the fewer temporal particles there are mediating the attraction. For example, imagine two objects suspended in mid-air near the Earth's surface that need to be pulled down, with the first having 1 unit of mass and the second having 5 units of mass. Imagine that a single man, representing Earth's gravitational attraction, can handle 1 unit of mass using all of his strength. The first object will be pulled down by a single man, whereas the other object will be pulled down by 5 men of equal size and strength. The result is that the objects are pulled down (or fall) at the same rate. The 1-unit and 5-unit masses also gravitationally pull on each other, but this effect is negligible because Earth's gravitational field is overwhelming compared to theirs.

Additionally, the 1- and 5-unit masses are pulling on Earth. However, a vastly more massive object such as Earth will pull more on an object of lesser mass in its gravitational field than the smaller object will pull on the planet. For example, a feather falling to the ground is of course being pulled by Earth, but Earth is also being pulled by the feather in the other direction. However, with its large mass and thus large gravitational field, Earth, in a sense, has all the men it needs to directly pull on the entire mass of the feather, but the feather, being so much less massive and thus having a smaller gravitational field, does not have all of the men needed to directly pull on the entire mass of the Earth—if a man could stand on a feather, of course. Thus, Earth pulls more on the feather than the feather pulls on Earth, but the feather does indeed pull on the Earth some, as much as it can.

The correlation between an object's mass and the amount of temporal particles involved in gravitational attraction not only causes such objects as feathers and bowling balls to fall down at the same rate toward Earth but also causes them to have the "sensation" of floating as they fall. This is because every part of the object—every part of its mass—is also pulled down at the same rate, so no strain develops within it.

Also, when a body falls toward another under the force of gravity, it accelerates as it moves. In TET, this occurs because the closer an object gets to the surface of the object it is falling toward, the more it enters into areas of increasing temporal-particle density, and the more tightly temporal particles are packed, the stronger the bonds they are able to have with each other and with an object's mass. Thus, as the object falls, the temporal particles, in effect, pull harder and harder on it, causing it to accelerate. Also, because of the increasing density and concomitant increasing interaction strength among the temporal particles, the closer two objects are, the stronger their gravitational attraction, and conversely, the farther apart they are, the weaker their gravitational attraction.

As in GR, space in TET (interior space) is curved where gravity is in effect. However, there is a slight difference in how the curvature is typically described in GR from how it is described in TET. In GR, the curvature is usually described as a spatial depression, like the curvature of a flexible sheet when a ball is placed onto it, with the surrounding areas of the sheet (space/space-time) being stretched in the direction of the curvature. In TET, the curvature is best described not by spatial depression but by spatial compression, caused by the influx of temporal particles. Whenever temporal particles congregate, they compress the space between them. For example, as temporal particles move toward Earth, forming its gravitational field, they compress the space surrounding the planet, curving the space as a result. As in the spatial-depression concept, the space surrounding the area of curvature brought about by spatial compression is stretched. Consider a loose-fitting sheet bound to the corners of a bed. If the fabric in the middle of the bed is gathered into a clump (compressed), the outer, surrounding areas of the sheet are stretched. Interestingly, the spatial-depression and spatial-compression scenarios look similar (*Figure 2.6*).

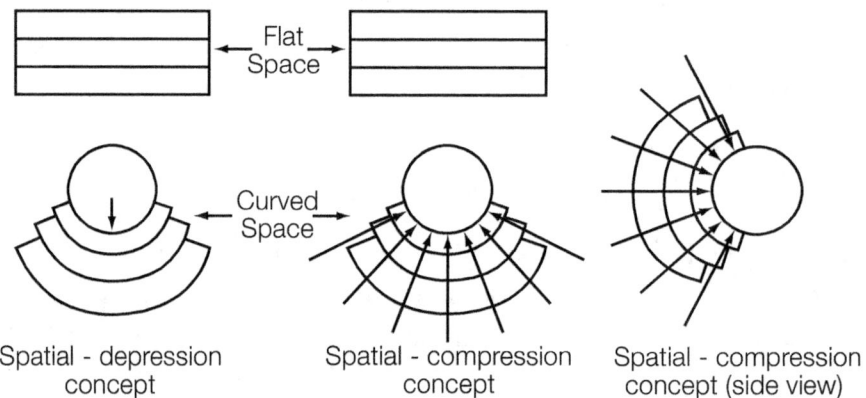

Figure 2.6 *The GR-related spatial depression and TET-related spatial compression concepts are visually similar. In GR, however, the body curves the space around it, whereas in TET, the space is curved by the movement of temporal particles toward the body.*

As the space surrounding a massive object is curved, so too is its gravitational field, as the field forms in spherical layers around it (*see Figure 2.3*). Because the gravitational field is time (temporal particles), it is correct even in TET to say that time is curved around the massive body. Thus, when space is curved, time is curved, and when time is curved, space is curved in both GR and TET. Note, however, that in TET, it is neither the curvature of space nor time that is causing gravity. Many have believed that gravity occurs because spacetime is curved, but in TET, it is the other way around. Space and time are curved because gravity—the flow of temporal particles toward matter and ultimately into exterior space—is occurring. It is a matter of swapping GR's cause and effect: The curvature of space and time does not cause gravity—gravity causes the curvature of space and time.

Gravitational Field Energy in GR and TET

In TET, temporal particles have a great deal of energy; indeed, they move faster than the speed of light. Thus, temporal particles, particularly t^+ particles, can be referred to as temporal energy, and this energy is of course gravitational field energy. There are correlations between GR and TET's version of this energy. Recall the TET idea that t^+ particles move toward matter particles and anchor themselves on their mass; this is equivalent to the GR concept of a gravitational field "coupling" to matter. When this occurs in GR, the energy associated with this coupling vanishes, which some have interpreted as a "problem." From the perspective of TET, this vanishing is related to the conversion of gravitational field energy from t^+ to t^- by matter and the subsequent absorption of t^- by exterior space. Because the energy is flowing ultimately into exterior space, of which we have no direct experience, it appears to simply vanish from our perspective. Thus, in TET, the vanishing of the energy is not a problem, but a natural and, indeed, necessary process in gravity.

Note that some of the vanishing energy in GR is specifically related to the matter particles themselves—that is, to the matter particles' mass and momentum. As is discussed in later sections, mass and momentum are also t^+ energy in TET, and thus are also subject to the same conversion and absorption processes that gravitational energy is. The mass and momentum of a matter particle essentially become part of the gravitational field energy, which is lost to exterior space. The gravitational field energy seems to subtract energy from the mass and momentum, as it itself is ultimately absorbed by exterior space. As is also discussed later, mass is continuously replenished as more t^+ particles move in, and thus, in most instances, mass appears constant over time. Momentum remains constant when matter travels through space at a constant velocity; as with mass, this momentum is continuously replenished as more t^+ particles move in to replace those consumed.

The fact that the energy of matter is not known to mysteriously vanish in GR unless it is coupled to gravitational field energy suggests that, as in TET, what is actually vanishing is gravitational field energy and this energy is subtracting from the matter's energy as the gravitational field energy itself vanishes. The apparent loss of energy during the coupling of a gravitational field to matter further suggests that this energy is continuously lost to the vacuum at the *local* level in that theory—in this case at the level of the matter. Indeed, except for a special mathematical case involving an infinitesimal region, energy is not conserved in general at the local level in GR.[7] In some cases, there is a loss of energy, such as the vanishing of gravitational field energy, mass, and momentum at the level of matter. And in other cases, there is a "creation" or seemingly spontaneous appearance of energy, such as at the level of an area of space called a white hole in the theory.[8] Although there is still some debate about the *global* conservation of energy in the theory—that is, the conservation of energy in the universe as a whole—GR does likely conserve energy in this manner.[7]

From the perspective of TET, the reason gravitational field energy, mass, and momentum are lost locally, at the level of matter, is because it is matter that is driving the conversion of t^+ to t^-. This is similar to the water in a tub leaving through a localized area, namely the drain. Note, however, that when it comes to the draining of the energy, the drain is actually space not matter. Again, matter just converts time so that it can be drained away. Space does the draining. In GR and TET, gravitational field energy, mass, and momentum are lost also in black holes—these are regions of space where a great deal of matter has been compressed into a very small volume (and is ultimately destroyed). Wherever there are particles of matter, such energy is disappearing, for inexplicable reasons in GR, but in TET, because t^+ is being converted to t^- by the matter and t^- is being absorbed by exterior space. In TET, the term *singularities* will be used to refer to the actual regions of space through which t^- is drained away. Thus, in this theory, singularities exist wherever matter exists, in normal space, as well as at the center of black holes, where they have usually been thought to only exist.

The emergence of energy from white holes in GR is akin to the process in TET in which t^+ particles are ejected from exterior space into interior space. From our perspective, living in interior space, energy seems to appear out of nowhere. The emergence of energy from white holes—which can also be referred to as time-reversed singularities—is an important, yet often overlooked, aspect of GR. Many have not taken white holes very seriously, considering them to be non-existent, a mathematical curiosity. It is a mistake to do this, however. The emergence of energy from white holes is an important process and is likely the mechanism responsible for the global conservation of energy in GR. In TET, as energy is lost through singularities (one type of localized area) in the vicinity of matter, it is restored through white holes (another type of localized area) where matter is scarce, such that, overall, energy is conserved. Thus, in TET, and likely also in GR, energy cannot be said to be conserved locally, but can be said to be conserved globally.

The idea of drains and spigots is useful in describing singularities and white holes. A drain is a localized area, and so is a spigot. A drain is an area through which water leaves a tub, and a spigot is an area through which water enters a tub. Singularities are like drains, and white holes are like spigots. If the tub has a volume of water in it, and the rate at which water leaves the tub through the drain matches the rate at which water enters the tub through the spigot, the volume of the water in the tub will not change. Thus, locally, in the area of either the spigot or drain, the volume of water is not conserved, as some water is entering the system in one area and leaving the system in the other. Globally, however, the volume of water is constant.

Although GR does not consider energy to physically leave our vacuum through black holes/singularities, and although white holes are not considered to be real in the theory, the activities of black holes and white holes do balance each other mathematically. That is, as energy disappears into black holes, energy spontaneously appears from white holes. There are some interpretations of GR in which white holes are considered to be the other side of black holes, such that the energy appearing from white holes is considered to be the same energy that disappeared into black holes.

General Relativity and Temporal Energy Theory

The point of the previous description of gravity by using TET concepts is not to suggest that there is anything wrong with GR. Instead, it is to suggest that perhaps GR is an approximation of TET. Many of the features of GR can be explained through TET concepts and in more detail, such as gravity, the connection between space and time, and the vanishing of energy when a gravitational field is coupled to matter. In both theories, time is central to gravity. In GR, time is considered to be a dimension akin to the three dimensions of space. In TET, time is considered to be both a process and a physical object—a uniform field of extremely small particles extending throughout the space in which we live (as well as the individual particles themselves). Because this field of temporal particles permeates every corner of our three-dimensional space, it does act like an extra, or fourth, dimension associated with that space. Thus, the dimension and field ideas are equivalent. Indeed, where gravity is in effect each is curved. Moreover, space is curved in both theories where gravity is operating.

However, the main takeaway message from this chapter regarding the potential relationship between GR and TET is that GR's description of gravity appears to rely on TET's t^+/t^- temporal-conversion process. As the next chapter will demonstrate, Quantum Mechanics (QM) is also likely an approximation of TET, but in contrast to GR's description of gravity, QM's description of the three other known fundamental forces (the electromagnetic, strong nuclear, and weak nuclear forces) appears to rely on TET's t^-/t^+ temporal-conversion process. And as stated in the Introduction, the possibility that GR and QM use equal, yet opposite temporal processes to describe their forces is likely the root of the so-called problem of time—the problem of the combined mathematical framework of GR and QM having no time. Simply add the t^+/t^- process to the t^-/t^+ process, and the obvious answer is zero. There appears, mathematically, to be no temporal process, no time, at work in the universe.

CHAPTER
3

QUANTUM MECHANICS

The three other important forces in the universe known to date, beyond gravity—namely the electromagnetic, strong nuclear, and weak nuclear forces—are described by Quantum Mechanics (QM). The electromagnetic force makes a magnet stick to a refrigerator door and an electrically charged balloon stick to a person's sweater. Actually, with the exception of gravity, all forces that we can directly experience as a push or pull are electromagnetic. These include frictional forces and drag forces, among others; such forces are brought about by the electromagnetic forces exerted by one atom on other atoms.[9] The other two fundamental forces—the strong nuclear and the weak nuclear forces—act over such short distances that we cannot directly experience them through our senses.[9] The weak nuclear force, or simply the weak force, causes one type of particle to change into another type. The strong nuclear force, or simply the strong force, holds the components of protons and neutrons together and is also thought to hold protons and neutrons to one another to form the nucleus of an atom (*Table 3.1*).

Table 3.1 *Forces of Quantum Mechanics*

Electromagnetism
Weak nuclear force
Strong nuclear force

QM describes these forces as being brought about by matter particles exchanging other particles—called *elementary bosons* (pronounced "boh-zonz"), *force carrier particles*, or simply *force carriers*—between them. Sometimes, these force carriers are undetectable, however, and are called *virtual particles*. Scientists say the reason they are sometimes undetectable is because at times their existence is extremely brief, in which case only their effects are directly measurable. Their existence is very brief whenever they are created through energy that is borrowed from the vacuum because this energy must be paid back very quickly. When sufficient additional energy is applied to matter particles, however, some of the force carriers that were virtual before can obtain some of this additional energy and become observable. Different force carriers produce different forces. The electromagnetic force is produced by the force carriers called photons; the weak force by W^+, W^-, and Z^0 bosons; and the strong force by gluons (pronounced "glue-onz").

One of the principal ideas in QM is that energy comes in discrete, little bundles called quanta, hence the term "Quantum" in "Quantum Mechanics." The force carriers are such bundles, as are the elementary matter particles, such as the electron. Thus, according to QM, two electrons, for example, which are tiny bundles of energy, exchange photons, other tiny bundles of energy, and in so doing create an electromagnetic force between them. Each electron is considered to absorb the photons that the other electron sent its way. A natural question is, Where do the force carriers come from? The answer is that they come from the vacuum—from space itself—through energy applied to the matter particles or through energy borrowed from the vacuum.

Table 3.2 shows the types of elementary matter particles in the universe and the force carriers they are said to exchange. Different matter particles cause the vacuum to produce different force carriers and thus experience different forces. There are 12 types of elementary matter particles, also called elementary fermions. Of these, six are called leptons, and the other six are called quarks. Three of the leptons are electrically charged. The other three are electrically neutral. The quarks have either a one third or two thirds fraction of the electric charge of the leptons. The charged leptons exchange photons to create the electromagnetic force. The quarks exchange gluons to create the strong force. All of the matter particles are capable of inducing the vacuum to create W or Z bosons and thus experiencing the weak force. The discussion that follows delves into the relationship between TET and the forces of QM.

Table 3.2 *Elementary Matter Particles and Their Associated Force Carriers*

Elementary Matter Particle			Associated Force Carriers	Force Produced
Name	Symbol	Electric Charge		
Electron	e^-	−1	Photons	Electromagnetism
Muon	μ^-	−1	Photons	Electromagnetism
Tauon	τ^-	−1	Photons	Electromagnetism
Up quark	u	+2/3	Gluons	Strong force
Down quark	d	−1/3	Gluons	Strong force
Charm quark	c	+2/3	Gluons	Strong force
Strange quark	s	−1/3	Gluons	Strong force
Top quark	t	+2/3	Gluons	Strong force
Bottom quark	b	−1/3	Gluons	Strong force
Electron neutrino	v_e	0	W or Z bosons	Weak force
Muon neutrino	v_μ	0	W or Z bosons	Weak force
Tauon neutrino	v_τ	0	W or Z bosons	Weak force
Plus quarks and other leptons			W or Z bosons	Weak force

(Leptons — Electron, Muon, Tauon; Quarks — Up, Down, Charm, Strange, Top, Bottom; Leptons — Electron neutrino, Muon neutrino, Tauon neutrino)

The Electromagnetic Force and Temporal Energy Theory

The electromagnetic force is so named because it was discovered that electric and magnetic forces are intimately linked. However, in describing this force, it is usually helpful to first discuss electric and magnetic phenomena separately.

The Electric Force

We typically learn very early that there is such a thing called *electric charge* (positive and negative) and that two electrically charged particles will exert a force on one another. Like charges repel each other (two positive charges or two negative charges), whereas unlike charges attract one another (a positive with a negative). According to QM, this force of attraction or repulsion occurs through the exchange of photons by the charged particles, as described above. However, another way of looking at this is that the attraction or repulsion occurs by way of what are called *electric field lines* associated with each

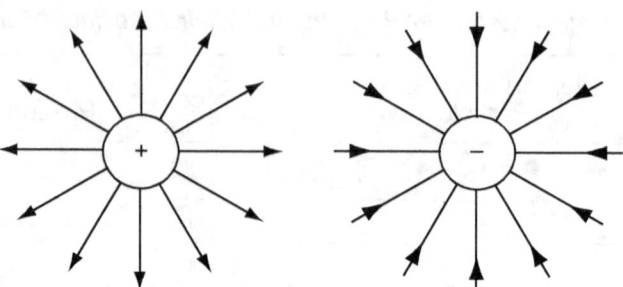

Figure 3.1 *An electric field can be represented by electric field lines emanating from a charged particle. Electric field lines are considered to extend away from positive charges and toward negative charges.*

charged particle. Electric field lines are considered to extend away from positive charges and toward negative charges (*Figure 3.1*). If the two particles are oppositely charged, the field lines as drawn out would appear to begin on the positively charged particle and end on the one with negative charge. If they are like charges, the field lines would still be present, but they would not be connected between the particles, as the two sets of lines would be pointing in opposite directions. A gap would appear between the particles indicating that they are repelling one another. In the case of a negatively charged particle, it may seem strange for the field lines to point toward the particle, considering that they are actually radiating outward from the particle just as the field lines of a positively charged particle are, but this is just a convention to distinguish positive from negative and help show the connection between oppositely charged particles (*Figures 3.2 and* 3.3).

This is the traditional way of describing the electric force and is usually the way that is initially taught to students. Michael Faraday was the first to describe these field lines, and he considered them to be real things. However, due to the advent of QM, which describes the force through the exchange of photons rather than through electric field lines, they fell out of favor with scientists. They are still typically used in such settings as introductory science classes because they are a useful tool for visualizing patterns in electric fields, but the field lines are not considered to be real things.[10] Although the concept of electric field remains, the idea of what constitutes the electric field and, consequently, what creates electric force has changed: electric field lines (according to Faraday, or the classical approach), photons (according to QM, the newer approach).

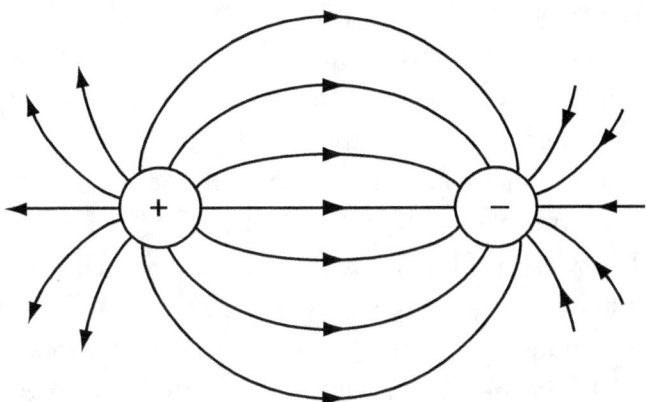

Figure 3.2 *The electric field lines between two oppositely charged particles connect with each other.*

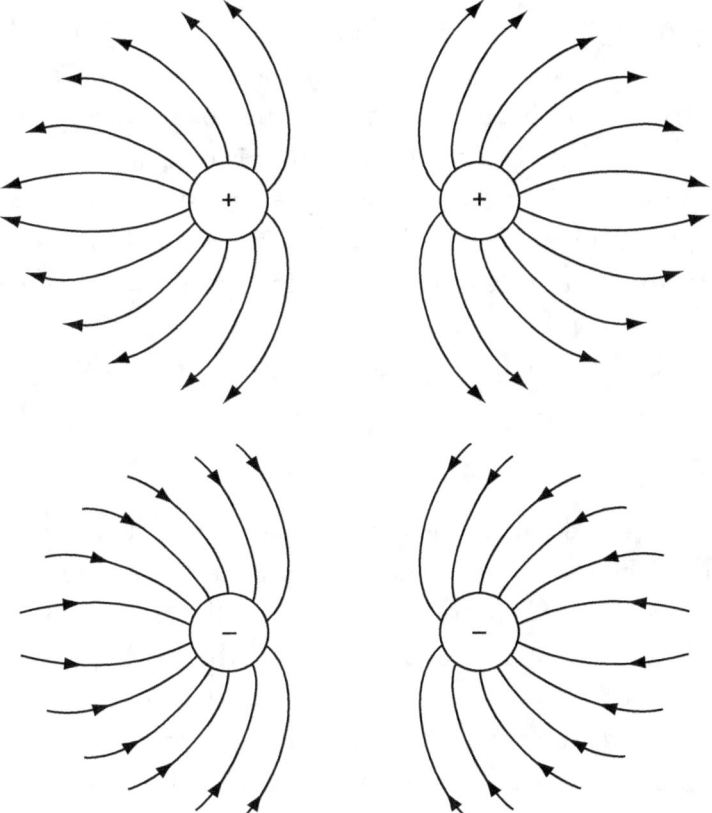

Figure 3.3 *The electric field lines between two positively charged particles or two negatively charged particles do not connect with each other. Instead, they bend away from each other, demonstrating a repulsive force between the charges.*

TET revisits the idea of electric field lines, but offers another explanation of how they may work. In TET, an electric field line should be thought of as a property of interior space and relates to its time. As noted earlier, exterior space, on which interior space is superimposed, temporally "breathes" in the t^-/t^+ direction. Interior space, on the other hand, breathes in both the t^+/t^- and t^-/t^+ directions. That is, unlike matter, which breathes in the t^+/t^- direction, and exterior space, which breathes in the t^-/t^+ direction, interior space breathes in both directions simultaneously and thus is temporally neutral. However, because matter exists only in interior space, it can be said that interior space does predominantly breathe in the t^+/t^- direction, but only by *proxy of matter*. (The reason exterior space breathes in the t^-/t^+ direction, while interior space breathes in both directions simultaneously will be discussed in chapter 9.)

Interestingly, the ability of interior space itself to breathe in the t^+/t^- and the t^-/t^+ directions can be separated out through a process called temporal polarization. By way of this process, a small area of interior space takes on a particular type of respiration, say t^+/t^-, with a sister area of equal size immediately next to it taking on the opposite type of respiration, in this instance t^-/t^+. Note that the temporal neutrality of interior space is maintained overall between the two sister areas of space. Temporal polarization can occur randomly in interior space wherever there are major spatial fluctuations, but it is principally brought about by electrically charged particles. These particles have the ability to temporally polarize the regions of interior space immediately surrounding them.

Thinking of a charged particle as a circle, the polarization occurs in a hub-and-spoke fashion—similar to Faraday's electric field lines—with the charged particle, such as an electron, being the hub and the areas of polarized space being the spokes. In the case of the electron, the initial polarized areas of interior space, the initial spokes, adopt the same type of temporal respiration as the charged particle itself, t^+/t^-. The initial polarized spokes cause equal amounts of interior space to be polarized oppositely—that is, in the t^-/t^+ direction—along imaginary lines set by the initial spokes. These are the sister areas of space, or sister lines, which again, maintain the temporal neutrality of interior space between them. The formation of the t^-/t^+ line induces the formation of another t^+/t^- line, which in turn induces the formation of yet another t^-/t^+ line, and so on (*Figure 3.4*). Each temporally polarized line of interior space—each t^+/t^- and t^-/t^+ line—is an electric field line, as well as the chain of them. The whole volume of these lines is the electron's electric field. It is important to remember, however, that an electric field line is really just a property of interior space itself. A charged particle's presence simply jump-starts the polarization process, with interior space taking over from there, adding more and more of these alternating polarized lines, with the phenomenon stretching as far as interior space itself stretches.

The "blue and yellow make green" idea can be used to help organize the concepts: Anything that breathes in the t^+/t^- direction will be considered blue. Anything that breathes in the t^-/t^+ direction will be considered yellow. And anything that breathes in both directions will be considered green. Therefore, the electric field lines breathing in the t^+/t^- direction are blue electric field lines. Those breathing in the t^-/t^+ direction are yellow electric field lines. Exterior space, which also breathes in the t^-/t^+ direction, is yellow space. Interior space, which breathes in both directions, is green space. An electron

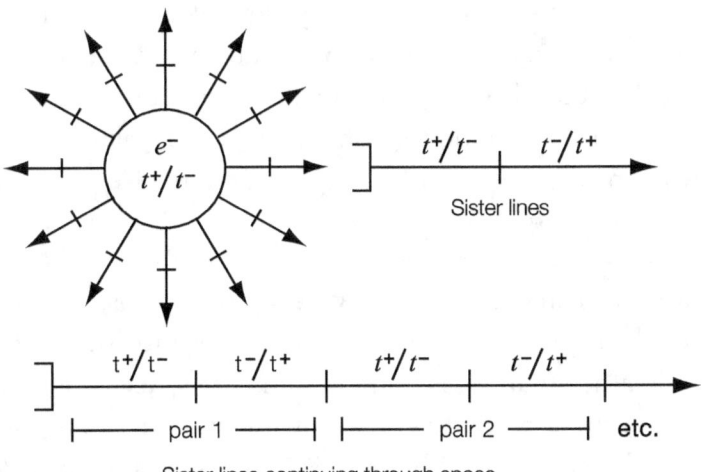

Figure 3.4 *In TET, an electrically charged particle temporally polarizes interior space, with one small area in the vicinity processing time in one direction (e.g., t^+/t^-) and a sister area processing time in the opposite direction (t^-/t^+). In the theory, lines of temporally polarized space are electric field lines. Once started, temporal polarization continues to occur, with temporally polarized lines of space stretching as far as interior space itself stretches.*

should be regarded as being like any one of the initial electric field lines emanating from it, except wrapped into a circle, and as it also breathes in the t^+/t^- direction, it is considered to be blue. As might be expected, temporal polarization and electric charge are one and the same. Continuing with the color scheme, anything that breathes in the t^+/t^- direction is blue and negatively charged. Anything that breathes in the t^-/t^+ direction is yellow and positively charged. And anything that breathes in both directions is green and electrically neutral (*Table* 3.3). Note, however, that there are particles that

Table 3.3 *TET Color Scheme*

Blue	Negative electric charge
	t^+/t^- respiration
Yellow	Positive electric charge
	t^-/t^+ respiration
Green	Electrically neutral
	t^+/t^- and t^-/t^+ respiration

have both blue and yellow elements. An excess of blue elements would make them negatively charged; an excess of yellow elements would make them positively charged. Equal amounts of blue and yellow elements would make them electrically neutral (being functionally green but not actually green). A proton is an example of a particle that has an excess of yellow elements, making it positively charged. A neutron has equal amounts of each and, hence, has no overall electric charge.

Note that even with an excess of yellow elements, protons, as well as neutrons, breathe overall in the t^+/t^- direction. This is because they exist in interior space and t^+, rather than t^-, is abundant in that space. Therefore, the blue elements within these particles effectively direct the particles' global respiration. Some of the t^- produced by the t^+/t^- processes aids the t^-/t^+ processes within these particles, considering that there is not much t^- available in interior space. However, overall, t^- is lost to exterior space. Thus, the statement above that anything that breathes in the t^+/t^- direction is blue and negatively charged still applies, in that protons and neutrons are, at the very least, partly blue and thus partly negative, from the perspective of TET.

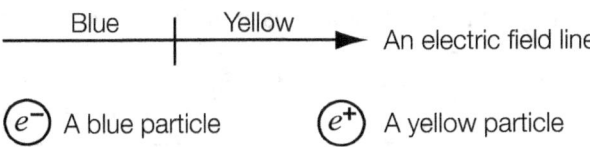

Figure 3.5 *In TET, the electron and positron are very much like a pair of blue and yellow sister electric field lines, except that they are separated and each is wrapped into a circle.*

Whereas the proton is a positively charged composite particle with blue and yellow elements, there is a particle that is positively charged and yellow all over—the positron. This is the sister particle of the electron, which is blue all over. The electron and positron are very much like a pair of blue and yellow sister electric field lines, except that they are separated, and as noted above for the electron, wrapped into a circle (*Figure* 3.5). Whereas the electron is called a matter particle, the positron is called an anti-matter particle. Anti-matter particles are associated with the same forces as their matter counterparts and are similar to their matter partners with the exception that some of their properties, like electric charge, are opposite (*Table* 3.4). One important exception is temporal respiration, however. All elementary or composite matter and anti-matter particles, with the exception of the electron and positron, breathe to some degree in both directions, having both blue and yellow and, in some cases, green elements. Because these particles exist in interior space, and because t^+ and not t^- is abundant in that space, overall these particles will breathe in the t^+/t^- direction, like the protons and neutrons described above, or at the very least, they will always be losing some t^- even if their respiration favors t^-/t^+. Although some of the t^- produced by their blue elements (or the blue side of their green elements) will aid their yellow elements (or the yellow side of their green elements), overall, t^- is lost to exterior space. The electron and positron are the only purely blue and purely yellow particles.

Note that a particle with blue and yellow elements will not necessarily process time in the t^+/t^- direction at an equal rate to its anti-matter sister. For example, the anti-proton is the proton's anti-matter counterpart. As the proton is positively charged, having an excess of yellow elements, the anti-proton is negatively charged, having an excess of blue elements. Both particles process time overall in the t^+/t^- direction, but the anti-proton does this more than the

Table 3.4 *Elementary Anti-Matter Particles and Their Associated Force Carriers*

Elementary Anti-Matter Particle			Associated Force Carriers	Force Produced
Name	Symbol	Electric Charge		
Positron	e^+	+1	Photons	Electromagnetism
Anti-muon	μ^+	+1	Photons	Electromagnetism
Anti-tauon	τ^+	+1	Photons	Electromagnetism
Anti-up quark	$\bar{u}$	−2/3	Gluons	Strong force
Anti-down quark	$\bar{d}$	+1/3	Gluons	Strong force
Anti-charm quark	$\bar{c}$	−2/3	Gluons	Strong force
Anti-strange quark	$\bar{s}$	+1/3	Gluons	Strong force
Anti-top quark	$\bar{t}$	−2/3	Gluons	Strong force
Anti-bottom quark	$\bar{b}$	+1/3	Gluons	Strong force
Anti-electron neutrino	$\bar{\nu}_e$	0	W or Z bosons	Weak force
Anti-muon neutrino	$\bar{\nu}_\mu$	0	W or Z bosons	Weak force
Anti-tauon neutrino	$\bar{\nu}_\tau$	0	W or Z bosons	Weak force
Plus anti-quarks and other anti-leptons			W or Z bosons	Weak force

(Left side row groupings: Anti-leptons — Positron, Anti-muon, Anti-tauon; Anti-quarks — the six anti-quarks; Anti-leptons — the three anti-neutrinos.)

proton because it has more blue elements and thus experiences more forward-in-time movement than its matter sister particle. Because of their t^+/t^- time, most anti-matter particles will participate in gravity in a similar way to matter particles. Positrons also participate in gravity, but their participation is a little more complicated because they experience no t^+/t^- time, being yellow all over and thus experiencing only t^-/t^+ time, unlike all other anti-matter particles. Gravitational interactions involving positrons are discussed in chapter 5.

Combining the hub-and-spoke model with the color scheme from above, an electron can be regarded as a blue circle with blue electric field lines (or blue lines) radiating from it that are connected to yellow electric field lines (or yellow lines), and so on. Although only elementary particles, such as electrons, are best described as circles, all particles with one unit of negative charge

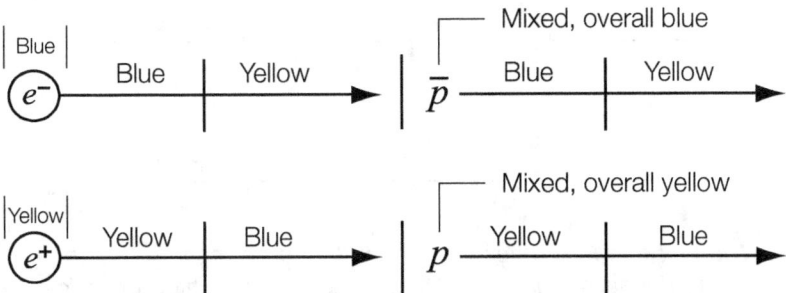

Figure 3.6 *Although an anti-proton (designated by a p with a bar over it) is a composite of other particles, it has one unit of negative charge and thus generates electric field lines in a similar fashion to electrons. Likewise, a proton (designated by a p) has one unit of positive charge and generates electric field lines in a similar fashion to positrons.*

generate electric field lines similarly to electrons. Thus, for instance, an anti-proton, a composite of several particles, but which like the electron has one unit of negative charge generates blue (t$^+$/t$^-$) lines initially, followed by yellow (t$^-$/t$^+$) lines, and so on (*Figure 3.6*).

Positively charged particles work similarly to negatively charged particles, with one key difference: Whereas interior space is initially polarized in the t$^+$/t$^-$ direction by negatively charged particles, it is initially polarized in the t$^-$/t$^+$ direction by positively charged particles. Beyond this, a unit of positive charge and a unit of negative charge work the same. The initial yellow (t$^-$/t$^+$) lines give rise to blue (t$^+$/t$^-$) lines, and so on, with the alternating blue and yellow lines stretching as far as interior space stretches and the collective volume of these lines being the charged particle's electric field (*Figure 3.6*). Note that, with the exception of the positron, positively charged particles will initially polarize interior space in the t$^-$/t$^+$ direction despite breathing globally in the t$^+$/t$^-$ direction. In conjunction with the environment of interior space, any and all blue elements in a positively charged particle direct its overall respiration. That is, as interior space is filled with t$^+$ particles, t$^+$/t$^-$ will be the dominant time process in a particle as long as it has a blue element in it. However, the continued presence of the particle's yellow elements will influence how the immediate area of interior space surrounding it will be initially polarized.

Although the members of a field line pair are equal (though opposite) to each other, the one that is first in the directional sense should be regarded as the

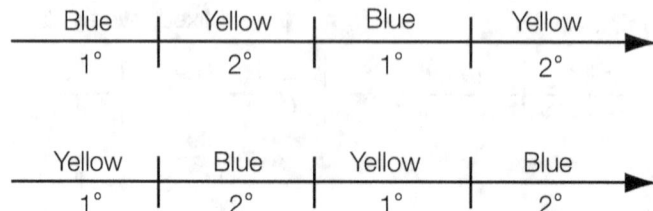

Figure 3.7 *The first line in a blue/yellow pair should be regarded as the primary line and the other as the secondary line. The first line may be blue or yellow.*

primary line and its sister as the secondary line, such that along a given chain of blue and yellow lines, there would be multiple primary lines and multiple secondary lines. In some instances, the blue (forward in time) lines are primary, whereas in other instances, the yellow (backward in time) lines are primary (*Figure* 3.7).

Interestingly, the equations describing an electric field, developed by James Maxwell, clearly have forward-in-time and backward-in-time solutions.[11] Unfortunately, the backward-in-time solutions have not been regarded as physically meaningful. They are an important part of TET, however, particularly in relation to the electric force. In TET, the electric force works as follows: A negatively charged particle starts its polarization of interior space with blue lines then yellow lines, creating a series of alternating lines thereafter, whereas a positively charged particle starts its polarization of interior space with yellow lines then blue lines, with it too creating a series of alternating lines thereafter, as shown in *Figure* 3.8, using two generic, electrically charged particles. Although technically a part of interior space, the field lines radiating from each of the charged particles should be thought of as extensions of them. For example, ignoring for the moment that ice floats in water, imagine a doughnut-shaped loop of ice resting in the middle of a tank of water, with the loop facing you for ease of comparison. Now imagine the ice causing some of the water surrounding it to also become frozen; to help with the analogy, consider the loop of ice and freezing water to form in a hub-and-spoke manner, with the rays of frozen water starting on the loop of ice and extending outward. The rays of frozen water emanating from the loop remain a part of the volume of liquid water from which they arose, but at the same time, they are extensions of the ice loop that caused their creation. In like manner, the temporally

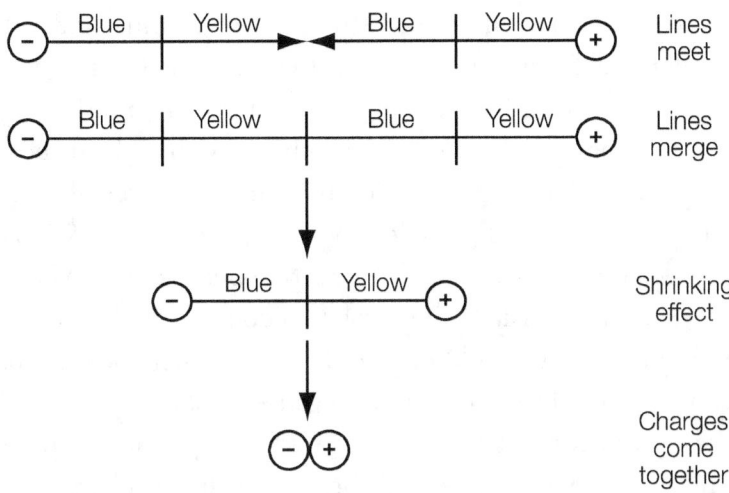

Figure 3.8 *In TET, the electric force of attraction comes about through the merging of the field lines between a negative and positive charge and the subsequent shrinking of the now common lines between them. (Only one common line is shown in the figure.)*

polarized lines emanating from a charged particle remain a part of interior space, but they are also a part of the charged particle that caused their creation.

This is an imperfect analogy because static electric field lines have no real substance to them, unlike the rays of ice, but it helps describe the idea that the field lines, even though they ultimately originate from interior space, are extensions of the charged particle, as the ice rays are extensions of the ice loop, even though the rays come from the volume of water into which the loop was placed. When a charged particle moves, its field lines move, as when the ice loop moves, its rays move. As the rays are stuck to the loop, so too are the field lines stuck to the charged particle. The temporal polarization of interior space should not be thought of as mere blue and yellow "stains on a sheet." Polarized field lines are, in a sense, raised up from the background, taking on a more independent existence from greater interior space.

In TET, the attractive force comes about because the opposing field lines merge, becoming one, and a sort of "shrinking effect" occurs within the now common field lines—the overall lines of alternating blue and yellow lines between the charged particles—pulling the two particles together (*Figure 3.8*). The speed at which they come together will depend on such factors as their mass. Note that the shrinking effect is not a "blue and yellow make green" cancellation. If this were the case, the field lines would just fizzle out between the particles and they would remain in their original positions. Rather, the shrinking effect has to do with the overlap and subsequent directional cancellation of the field lines. To understand this, recall that field lines are extensions of the charged particles. When the field lines of two oppositely charged particles meet, they are able to overlap because they share the same pattern. For example, consider the arrowless common line between the two charges in *Figure* 3.8. This field line could just as easily belong solely to the positively charged particle as it could the negatively charged particle; the pattern is the same. Actually, the arrowless image is slightly more accurate because once their field lines have merged, neither the positively charged particle nor the negatively charged particle can claim true ownership of the now common line. However, arrows are important to indicate that the two particles are actually radiating field lines in opposite directions. In merging with each other,

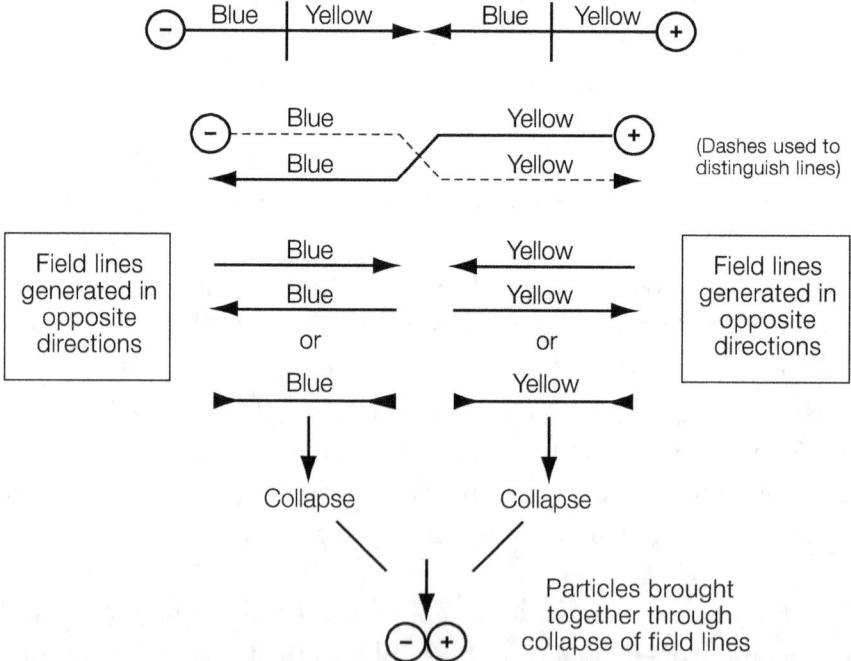

Figure 3.9 *When the field lines emanating from oppositely charged particles merge, the secondary lines of one particle overlap with the primary lines of the other particle. Because the overlapping lines are polarized the same but are being generated in opposite directions, they collapse or disappear.*

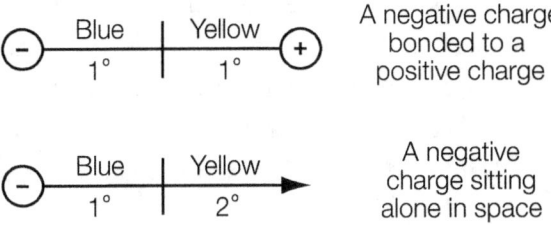

Figure 3.10 *When showing two particles electrically bonded to each other in TET, it is helpful to show a line of force between them and limit that line to just the particles' initial primary lines, ignoring the secondary lines, which should only be included when the particles are considered to be sitting alone in space.*

the secondary lines of one particle overlap with the primary lines of the other particle. Because the overlapping lines are polarized the same but are being generated in opposite directions, they collapse or disappear (*Figure 3.9*).

For example, consider the overlapping blue lines from *Figure 3.9*. The negatively charged particle is generating its line from left to right, but the positively charged particle is generating its line from right to left. The result is that the merged line goes nowhere; it collapses to zero. And this happens for every blue and yellow line in the common field line. The result of this is that the common field line itself collapses or shrinks, and the electrically charged particles of which those field lines are a part, are pulled together. Note that technically, there are no longer two blue or two yellow lines overlapping once the merge has occurred, although schematically it can be viewed that way to understand the dynamics that are occurring. When two lines overlap, they become one line. This is because only a *single* area of space is actually being operated on. That area of space is simply being blue polarized (or yellow polarized) in opposite directions, leading to zero polarization and collapse.

When showing two particles electrically bonded to each other in TET, it is helpful to show a line of force between them and limit that line to just the particles' initial primary lines, ignoring the secondary lines, which should only be included when the particles are considered to be sitting alone in space (*Figure 3.10*). You can think of the two primary lines as "getting in the way" of each particle's secondary line. Once the particles are separated, the secondary line of each particle reappears. In the case of an electric bond, the primary line of one particle looks like the secondary line of the other. The choice of how many primary-secondary pairs to show for a lone particle may depend on what is being demonstrated, but at least one pair should always be shown for such a particle when field line dynamics are being considered.

Note that when multiple field lines are considered, the curvature of those lines when the charged particles are close to each other should also be considered, as shown in *Figure 3.2*. In TET, the curvature of the field lines occurs because space is curved. Recall from the previous chapter that temporal particles surround sources of energy, and in so doing, compress the space they converge on, which gives that space a curved nature. When a charged particle with uniform mass sits alone in space, the curvature of the space surrounding it is also uniform. All of the space immediately surrounding the particle is just pushed back toward the particle, such that there is no curvature in the field lines radiating from the particle. (However, because of the spatial compression, the blue and yellow lines closest to the particle are shorter than those further out [*Figure 3.11*]).

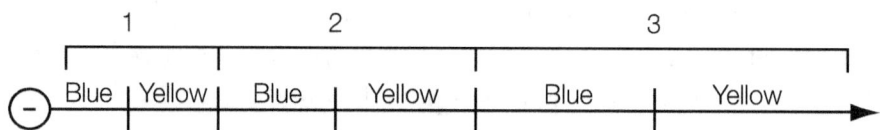

Figure 3.11 *As shown in Figure 2.3, temporal particles try to get as close to a source of energy as possible, with their density steadily increasing toward the surface of the object. Because of this, there is more spatial compression closer to the object. When the object is an electrically charged particle, the blue and yellow lines closest to the particle become shorter than those further out due to the varying spatial compression around the source particle. Even those in a sister pair may differ. That is, the primary line may be more compressed than the secondary line in the pair.*

When two electrically charged particles come together, however, the space between them is curved more than the space on either of their far sides, because there is much more energy between them. For instance, if particle A has 20 units of mass and particle B has 20, there would be 10 units of mass on the far side of particle A, 10 on the far side of particle B, and 20 between them—10 from particle A and 10 from particle B—with A and B considered to be uniform masses. Thus, there is more energy and spatial curvature between them. Because the temporal particles responding to the energy in the middle can only enter from the top or bottom of that area when the particles are close to each other, the space in the middle is mostly compressed from the top and bottom, as opposed to from the sides. As a result, the field lines in the middle are bent slightly downward from the top and upward from bottom, such that in some cases, those emanating from a particle become parallel or nearly parallel, particularly those in the deep middle of the area. This curvature of the field lines emanating from the charged particles helps the lines link up more than if they were in a straighter formation. That is, without the curvature, the field lines would be more crisscrossed than linked. Also, the curvature in the middle of the system will distort the shape of the field lines on the far sides of the particles as well—that is, the field lines on the far side of each particle will be more curved than if each particle sat alone in space.

With electric repulsion, which requires either two negatively charged particles or two positively charged particles, the "faces" of the opposing field lines are polarized the same. That is, two negatively charged particles will always have their yellow lines facing each other, and two positively charged particles will always have their blue lines facing each other. This prevents the polarized lines from overlapping and canceling each other out. Instead, they add up, pushing the charged particles away from each other (*Figure 3.12*).

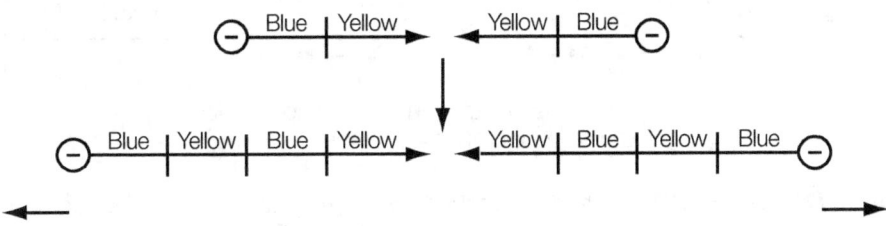

Figure 3.12 *In electric repulsion, the electric field lines between two identically charged particles cannot merge. Instead, more blue and yellow field lines are continuously generated, pushing the charged particles apart. (For simplicity, the spatial compression of the lines is not shown. See Figure 3.11)*

Although repelling field lines do not overlap, the "overlap method" described above for electric attraction can be used to view electric repulsion better: Consider two generic, negatively charged particles. As with electric attraction, consider the secondary line of each particle to overlap with the primary line of the other particle. In electric attraction, the merged primary and secondary lines had the same type of polarization (blue or yellow); they were just being generated in opposite directions, leading to collapse. In electric repulsion, the merged primary and secondary lines have opposite polarizations. One particle is trying to make the given area of space blue in one direction, and the other is trying to make it yellow in the other direction. Therefore, they cannot share any area of space between them. Because they continue to make polarized lines but those lines cannot cancel each other out, the lines propel the particles away from each other (*Figure 3.13*).

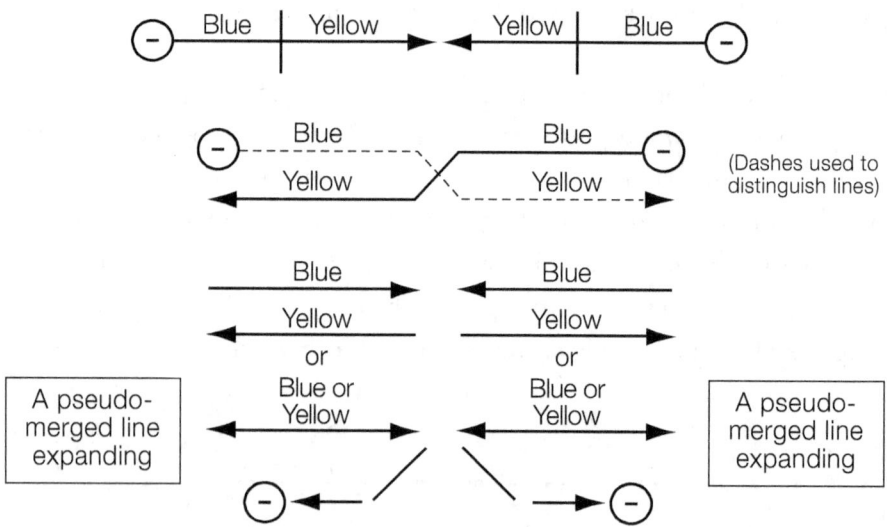

Particles driven apart through continuous addition
of more field lines between them

Figure 3.13 *The "overlap method" described for electric attraction can be used to view electric repulsion, using a pseudo-merging of the field lines. In electric repulsion, the merged primary and secondary lines have opposite polarizations. One particle is trying to make the given area of space blue in one direction, and the other is trying to make it yellow in the other direction. Therefore, they cannot share any area of space. Because they continue to make polarized lines but those lines cannot cancel each other out, the lines propel the charged particles away from each other.*

At first glance, using the overlap method, a "blue and yellow make green" change seems like it should occur, leading to a collapse of the field lines, but this is just an artifact of the method, because there is no real overlap occurring in electric repulsion. Whenever the overlap method is used, oppositely polarized lines generated in opposite directions represents repulsion, whereas identically polarized lines generated in opposite directions represents attraction. For the sake of completeness, oppositely polarized lines generated in the same direction *would* reduce to green if they could overlap, although turning green does not equal collapse. In collapse, the field lines literally shrink. In turning green, they would just disappear. Identically polarized lines generated in the same direction would still overlap to create a single field line, but the overlap would not lead to collapse, as collapse results from their being generated in opposite directions (*Figure 3.14*).

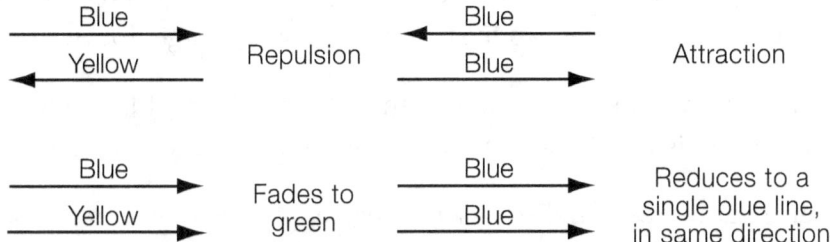

Figure 3.14 *The outcomes of interacting blue and yellow lines, depending on the colors of the lines and the directions in which they are generated.*

Faraday's electric field lines and their roles in electric attraction and repulsion are very likely just an approximation of these lines of alternating blue and yellow, forward-in-time and backward-in-time, polarized lines and their interactions. Faraday saw the field lines, and Maxwell saw the forward-in-time and backward-in-time nature of the electric force. TET's description of the electric force unites these two ideas.

Maintenance of Electric Charge and Field Lines

In the discussion above, it was stated that temporal polarization and electric charge are the same thing. If something is temporally polarized in the t^+/t^- direction, it is negatively charged (and blue in the color scheme), and if something is temporally polarized in the t^-/t^+ direction, it is positively charged (and yellow in the color scheme). A distinction should be made at this point, however, between the capacity to process time in a particular direction and the actual act of doing so. Temporal polarization concerns the capacity to process time unidirectionally, but if temporal respiration does not actually occur, the blue or yellow polarization can be lost. That is, objects can suffocate temporally if they do not get the type of time they need. Temporal suffocation would simply turn something that is blue or yellow into something green.

For example, an electron, a blue particle, takes in t^+ particles, converts them to t^0 first and t^- second, and then emits the t^- particles. If for some reason the electron could not perform this operation, its blueness would begin to fade. It would turn first bluish green (in the color scheme, not literally), and if there were still a problem, perhaps to true green. By turning bluish green, it gains the ability to maintain some polarization under suffocating conditions (that is, there being few t^+ particles around), and thus some charge, for two reasons:

1. A bluish-green particle does not need as much t^+ as a fully blue particle because there is less "blue," less polarization, to have to maintain.

2. By being partly green, it also, to some degree, processes time in the opposite direction, t^-/t^+, which provides it with additional t^+ it can use to maintain its slight blue polarization. That is, it does not throw much away, being able to use its own byproducts.

When a particle is true green, it processes time equally in both directions, sometimes using its own byproducts in future respiration cycles (t^+ for t^+/t^- and t^- for t^-/t^+), as described above. Note that "true green"/temporal neutrality should not be taken too literally. Something that is true green, such as interior space itself, will fluctuate slightly between its blue and yellow natures. Its "true greenness" is an average of these minor fluctuations.

It would be unusual for an electron or any other QM particle, except for the positron, to be in suffocating conditions, however. All of the particles exist in interior space, and as this space is filled with t^+ particles, electrons have all that they need. And all matter and anti-matter particles except the electron and positron breathe to some degree in both directions. This, combined with the idea that anything that processes time in both directions can use the by-products of one respiration mode for the other, enables these particles to be stable temporally. (Some may be unstable for different reasons.)

The positron, however, can suffocate, though this is not instantaneous. The particle can extract from interior space some of the t⁻ particles it needs to maintain its yellow polarization, because interior space processes time in both directions, so there is always some t⁻ in interior space at any given moment. The problem is that the positron is fighting for that t⁻ with all of exterior space, which uses t⁻ to maintain its own yellow polarization. Imagine a small flower planted next to a tree. The tree's larger roots soak up most of the water, leaving little for the flower's smaller roots. Temporal suffocation is likely the reason positrons do not exist in abundance today. Many in the early universe would have turned yellowish green quickly.

The same fate of the positron would befall the yellow components of electric field lines if they were not partnered with their blue sisters. Again, although some t⁻ can be extracted from interior space, anything that wants it must, in a sense, battle with the behemoth that is exterior space for it. The advantage of being in close proximity to the blue lines is that there are many "free" t⁻ particles, the byproducts of the blue lines' t⁺/t⁻ temporal-respiration process. Of course, exterior space soaks up most of these t⁻ particles also, but there is always a steady stream from the blue lines, so the yellow lines have what they need. Using the flower and tree analogy again, it is like the flower being steadily sprinkled with water. The tree may still soak up most of the water, but the steady stream in the flower's immediate vicinity helps the flower to survive (*Figure 3.15*).

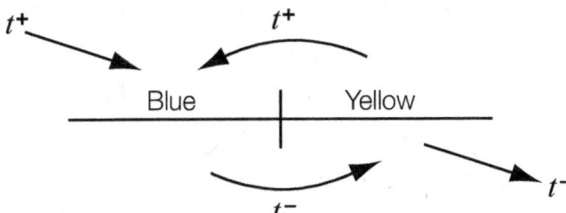

Figure 3.15 *The yellow components of electric field lines obtain the t⁻ they need through that produced by their blue sisters.*

Also helping in the maintenance of electric charge are the forces of attraction and repulsion between temporal particles and electric field lines and similar elements in QM particles, based on the spatiotemporal charge on the temporal particles and the temporal nature of the elements. Temporal particles form bonds with each other regardless of their charge, and all temporal particles are attracted to all forms of energy. However, temporally polarized elements do recognize the charge on temporal particles and are selective about the type of temporal particle they will let get close to them. Blue elements attract and form bonds with t^+ particles and repel t^- particles. Yellow elements attract and form bonds with t^- particles and repel t^+ particles. Thus, for instance, t^+ particles will constantly try to get close to a positron, a fully yellow particle (indeed the only fully yellow particle), due to its energy, but the positron will keep those particles at bay, allowing only t^- particles to be in its immediate vicinity.

This activity helps ensure that polarized elements in field lines and particles get the type of time that they need for their survival. Again, however, the electron is the only fully blue particle, and the positron is the only fully yellow particle. All other QM particles breathe to some degree in both directions, and so attract and form bonds with both types of temporal particles and thus have both types close to them. Note, however, that the blue elements in these particles, as well as the blue components of electric field lines, continue to repel t^- particles. The yellow elements of all mixed systems must grab the t^- their blue partners produce before the t^- is ejected too far from the systems. Green elements in particles have an equal or near-equal affinity for both t^+ and t^-. Bluish-green elements repel t^- very weakly—they will repel t^- only once their yellow "half" has what it needs in a given moment. Yellowish-green elements repel t^+ very weakly—they will repel t^+ only once their blue "half" has what it needs in a given moment. Being true green, interior space has an equal affinity for both types of particles; it does not repel either type. However, being fully yellow, exterior space has a stronger affinity for t^- particles than does interior space. Exterior space strongly repels t^+ particles. The t^0 particle is a transient state between t^+ and t^-; as a consequence, it has qualities of both.

The Magnetic Force

In the above description of the electric force, all of the electrically charged particles were essentially stationary in space. When electrically charged particles move, they experience another force: the magnetic force or magnetism—typically defined as an attractive or repulsive force between two electrically charged particles that are moving through space. This is in addition to the attractive or repulsive force those particles feel due to their electric charges. In fact, depending on circumstances, the magnetic force can overshadow the electric force. For instance, two electrons sitting in space would normally repel each other, but if they are moving alongside each other, they will attract one another magnetically; that is, the attractive magnetic force between them can override the repulsive electric force (*Figure 3.16*).

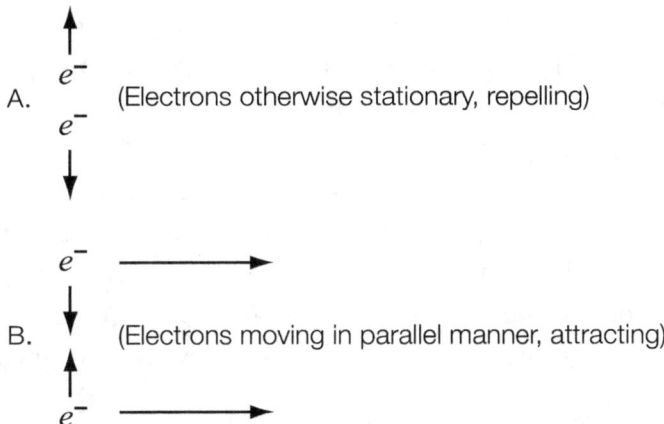

Figure 3.16 *Two electrons sitting in space would normally repel each other (A), but if they are quickly moving alongside each other, they will attract one another magnetically (B).*

Before the discussion of magnetism continues, it should be mentioned that from the standpoint of TET, everything that moves is capable of being magnetic at least to some degree, not just electrically charged particles. One of the main reasons electrically charged particles, particularly small ones like electrons, tend to stand out is because they typically move very, very quickly, amplifying the effects that really all particles feel when they move. From the perspective of TET, magnetism is caused nearly by the same phenomena that cause gravity. (As in the previous chapter, the focus will be on matter, but with the exception of the positron, the following list is identical for anti-matter. Magnetism from the standpoint of the positron is discussed in chapter 5.)

1. The attraction of t^+ particles to particles of matter, or more precisely, the attraction of t^+ particles to energy, including the energy of other t^+ particles;

2. The bonds t^+ particles form with each other and with a matter particle's mass—the more compact the temporal particles are, the stronger the bonds between them and also the mass of the matter particle they are connected to;

3. The conversion of t^+ to t^- by matter;

4. The absorption of t^- by exterior space.

In other words, gravity and magnetism are nearly the same force in TET. They each concern the flowing of temporal particles toward energy and ultimately their disappearance into the fabric of space itself—out of interior space and into exterior space. Recall that motion itself is a form of energy. Thus, t^+ particles being attracted to all forms of energy are attracted to this motion, swarming around moving objects the same way they swarm around massive objects in general. Motion is energy. Mass is energy. Temporal particles respond to both. Basically, when temporal particles respond to mass they cause gravity; when they respond to motion, they cause magnetism.

The principal difference between gravity and magnetism relates to the order of importance of the four items listed above. That is, while all four are generally important in both forces to produce their full effects, items three and four are the primary phenomena of gravity, while items one and two play more secondary roles in that force. In magnetism, it is the reverse: Items one and two have the primary roles, while items three and four are more secondary in that force. The conversion of t^+ to t^- by matter and the subsequent absorption of t^- by exterior space are the driving force of gravity. The attraction of t^+ particles to energy and the bonds the particles form just set the stage for gravity to occur. The dynamics of magnetism are born mostly from the attraction of t^+ to energy and the bonds the particles make, but the transformation of

t^+ to t^- and the absorption of t^- by exterior space help produce full magnetic effects.

Magnetism actually comes into play when there are two objects moving parallel or anti-parallel to each other in an area of space and is brought about through the interactions between the temporal particles that converge on the scene in response to the motion—these temporal particles compose the objects' magnetic fields. Basically, if the objects are moving in a parallel manner, the temporal particles between them are able to bond with each other and thus pull themselves and the objects to which they are bonded closer to one another, with temporal respiration also playing a role in bringing the objects closer together, as it does in gravity. If their movements are anti-parallel, bonds will form between the temporal particles but then break as the objects pass each other. The energy released from the breaking of the temporal bonds will cause the objects to fly apart faster. Also, when two objects move away from each other, instead of temporal respiration helping them come together, it aids in their moving apart, in that the t^- particles' contribution to the bonds fades or weakens as they are absorbed by exterior space, helping the bonds they are a part of to break that much more easily (*Figure 3.17*). The following examples help illustrate magnetic attraction and repulsion from the standpoint of TET concepts.

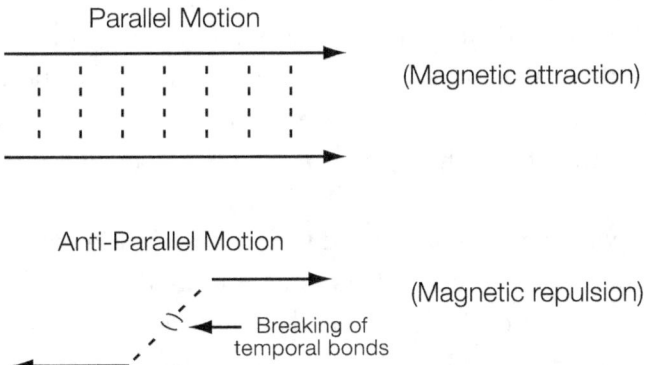

Figure 3.17 *When two objects are moving in a parallel manner, the temporal particles between them are able to bond with each other and thus pull themselves and the objects to which they are bonded closer to one another. If their movements are anti-parallel, bonds will form between the temporal particles but then break as the objects pass each other. The energy released from the breaking of the temporal bonds will cause the objects to fly apart faster.*

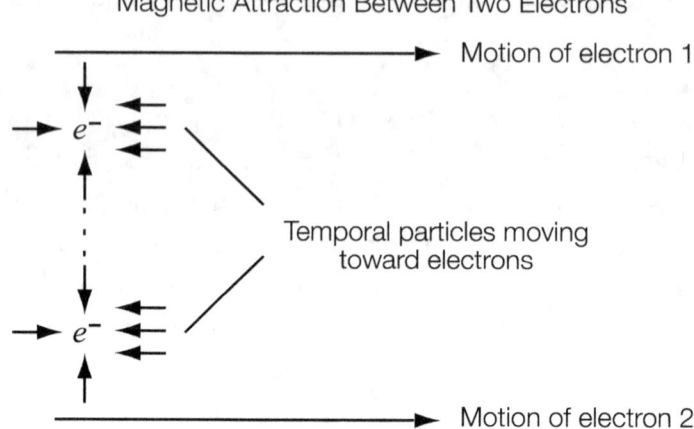

Figure 3.18 *The TET representation of two electrons moving parallel to each other through space and the bonds among the temporal particles between them, causing magnetic attraction.*

<u>Two Moving Electrons</u>

Consider two electrons moving parallel to each other through space. Note that temporal particles are approaching the moving electrons from all directions, but principally from the front. This makes intuitive sense because if temporal particles are attracted to motion, it is reasonable that they would be most abundant in the direction an object is moving, in the direct line of that motion, meeting it head-on. Recall that temporal particles move faster than the speed of light. Thus, their response to the movement of objects like electrons is extremely fast, and again, they are also attracted to each other's energy.

All of the temporal particles associated with each electron commingle and form bonds with each other—those of its magnetic field and those of its gravitational field—such that each electron and all of its associated temporal particles form a single system. As the two electron systems move through space, their respective temporal particles also commingle and form bonds with one another, which causes the two systems to move toward each other. As they move closer, the temporal particles form stronger bonds, substantially linking the two systems together, causing the systems to move even closer. Also, as with gravity, temporal respiration occurs in the two electrons causing a slight tension in the common field of temporal particles, as the electrons consume the temporal particles closest to them. The tension in the common field combined with continued respiration also causes the electrons to come together. These events, from the standpoint of TET, are what bring about magnetic attraction (*Figure 3.18*).

Now consider the two electrons moving anti-parallel to each other, that is, along two imaginary parallel lines but in opposite directions, passing each other in the process. This would cause a magnetic repulsion between the two electrons because the temporal particles of the two systems are moving in opposite directions. As the electrons pass each other, the bonds linking the two systems become strained and eventually snap, hurdling the electrons away from each other. Temporal respiration continues in the two electrons, producing t⁻ particles. As they are absorbed by exterior space, their contributions to the temporal bonds in the magnetic fields begin to fade or weaken, making it easier for the bonds to be broken (*Figure 3.19*).

When it comes to magnetism it takes two to "tango." Magnetism is an inter-action between the temporal particles responding to the movement of two objects as they move parallel or anti-parallel to each other. If one object were

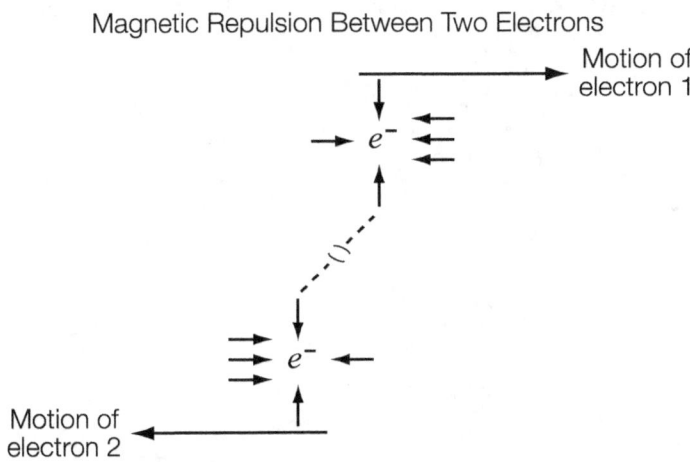

Figure 3.19 *The TET representation of two electrons moving anti-parallel to each other through space and the breaking of the bonds among the temporal particles between them, causing magnetic repulsion.*

moving and one were sitting still, magnetism in its truest sense would not occur. At best, as the moving object passed the stationary object, you might observe (1) a full gravitational attraction, (2) a slight attractive force—being somewhat of a gray area between a gravitational attraction and a magnetic attraction—(3) a slightly repulsive force, or (4) nothing at all. The outcome depends on such factors as the objects' comparative energy levels (particularly their mass), the distribution of mass within them, and perhaps even their shapes.

Two Current-Carrying Wires

The same principles described above are in effect when electrons move in conducting wires, that is, when there is an electrical current. This is because the two situations are essentially the same, except that in the above discussion, the electrons were moving through empty space, but here they are moving within the confines of wires. Thus, two wires lying next to each other with electrical currents running through them will be attracted to one another if those currents are parallel. If the electrical currents are anti-parallel, the two wires will repel each other (*Figure 3.20*).

The situation is nearly the same when the current-carrying wires are placed in loops. The only difference is that, when placed in loops, oppositely moving currents attract, whereas currents moving in the same direction repel one

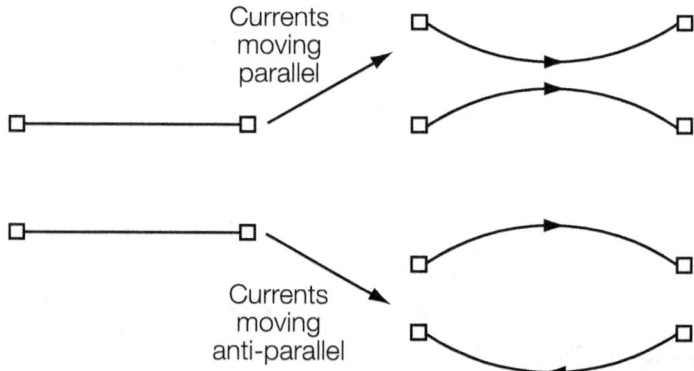

Figure 3.20 *Two wires lying next to each other with electrical currents running through them will be attracted to one another if those currents are parallel. If the electrical currents are anti-parallel, the two wires will repel each other.*

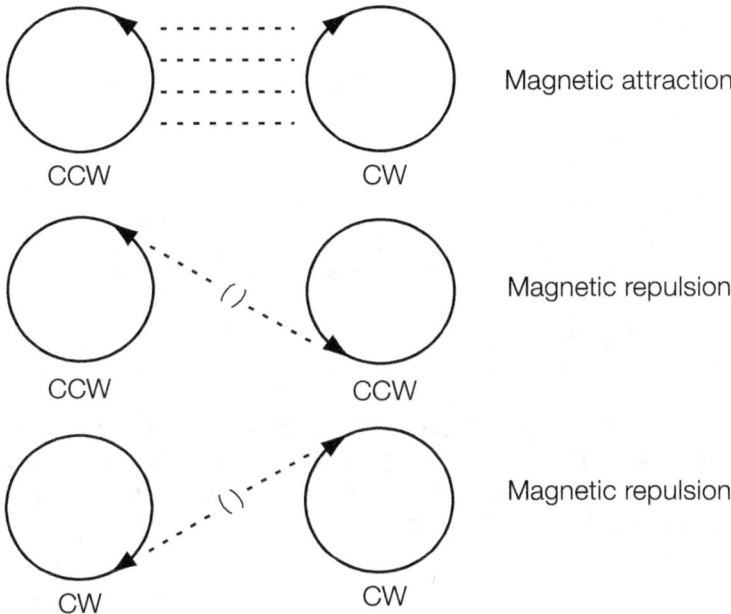

Figure 3.21 *If the wires are placed into loops and one current is revolving clockwise (CW) and the other counterclockwise (CCW), the wires will attract one another. If both are moving clockwise or if both are moving counterclockwise, they will repel one another.*

another (similar to electric charges, with opposite charges attracting and like charges repelling each other). The direction it is moving is determined by the direction it is moving while you are facing it. If one current is revolving clockwise and the other counterclockwise, the wires will attract one another. If both are moving clockwise or if both are moving counterclockwise, they will repel one another (*Figure 3.21*). Note that current-carrying wires placed in loops are akin to bar magnets. If you are looking at the loop and the current is moving in a counterclockwise manner, that is like looking at the north pole of a bar magnet. If you are looking at the loop and the current is moving in a clockwise manner, that is like looking at the south pole of a bar magnet (*Figure 3.22*). When two north poles or two south poles meet, they will repel one another, because this is like two loops of counterclockwise-moving currents (in the case of two north poles) or two clockwise-moving currents (in the case of two south poles) interacting with each other. When a north pole meets a south pole, they will attract each other, because this is like clockwise- and counterclockwise-moving currents interacting (see *Figure 3.21*).

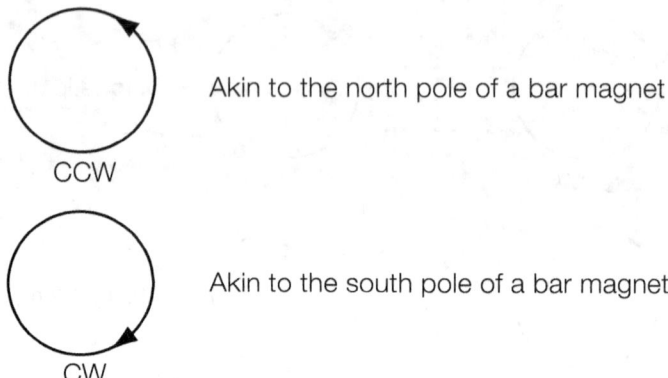

Figure 3.22 *Current-carrying wires placed in loops are akin to bar magnets. If you are looking at the loop and the current is moving in a counterclockwise manner, that is like looking at the north pole of a bar magnet. If you are looking at the loop and the current is moving in a clockwise manner, that is like looking at the south pole of a bar magnet.*

Bar magnets and current-carrying loops of wire are nearly indistinguishable from a purely magnetic standpoint. The reason for this in TET is that the magnetic nature of bar magnets comes from little current-like loops within the material—actually within the electrons of the material. The electron is the fundamental unit of magnetism in a bar magnet, and every electron has inside of it something akin to a little loop of electrical current, which scientists call its spin. In TET, it is actually a loop of spinning space (interior space) (*Figure* 3.23). Like a loop of electrical current, the spin of an electron develops a magnetic field around it. Recall the earlier statement that in TET, anything that moves is capable of being magnetic: This includes space, as well.

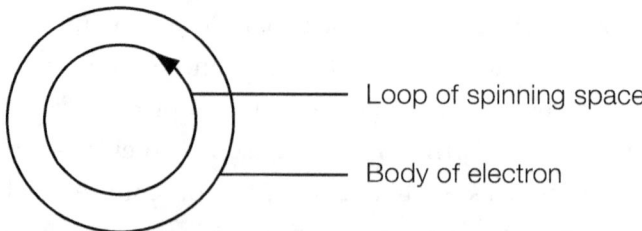

Figure 3.23 *The electron is the fundamental unit of magnetism in a bar magnet. From the perspective of TET, every electron has a loop of spinning space inside of it, called its spin, which develops a magnetic field.*

From a purely qualitative standpoint, the magnetic field created by a current-carrying loop is indistinguishable from a magnetic field created by the spin of an electron. Quantitatively, it usually takes the collective magnetic strength of many electrons, in for instance a simple macroscopic-sized bar magnet, to match the magnetic strength of a single, current-carrying loop of wire. Note that the electrons must be arranged in a particular way in a material for their magnetic fields to add up well enough to create macroscopic magnetic effects. The principal idea, however, is that the magnetism of bar magnets originates from loops of spinning space inside electrons from the TET perspective, and this is the reason bar magnets and macroscopic loops of electrical current behave similarly. They are similar phenomena (spinning loops). Note that electron spin is below the level of friction—and a body in motion remains in motion unless a force is applied to stop it. It is below the level of friction because it is below the level of the surface or "body" of the electron. It is internal to the particle. The north of bar magnets comes from the counterclockwise spin of electrons, and the south of bar magnets comes from the clockwise spin of electrons in the magnet.

Magnetism versus Gravity

A natural question is, Why is magnetism (attractive and repulsive) so much stronger than gravity? Refrigerator magnets stick to refrigerators by way of magnetic attraction rather than fall to the floor under the force of gravity. Through magnetic repulsion, some objects can be made to float in thin air, again defying gravity. From the perspective of TET, it is not that gravity is weaker _than_ magnetism, but rather that gravity is weakened _by_ magnetism. Consider a refrigerator and a refrigerator magnet. Deep inside the refrigerator are electrons that are arranged in such a way that their spins give the unit a magnetic nature. The refrigerator magnet also has electrons arranged in such a way that their spins bestow it with a magnetic nature. As noted above, the spin of an electron is a loop of spinning space with its associated magnetic field. Thus, in the attraction between the refrigerator and magnet, the collective spin of the electrons in the refrigerator (spinning for example clockwise) interact with the collective spin of the electrons in the magnet (spinning for example counterclockwise), leading to magnetic attraction between them.

Gravity is weakened because as the temporal particles associated with the magnetic fields spin around, the bonds they have with the temporal particles of Earth's gravitational field are constantly broken. Thus, the planet's gravitational field never gets a good grasp on the refrigerator magnet's magnetic field. In contrast, the bonds between the temporal particles composing the refrigerator magnet's magnetic field and the temporal particles composing the refrigerator's magnetic field largely remain intact during the interaction and, because of this, are stronger than the bonds with the Earth's gravitational field, leading to magnetic attraction overcoming gravity (*Figure 3.24*).

Gravity is also weakened during magnetic repulsion for the same reasons. Consider an object magnetically levitating above another object on a table due to a repulsive magnetic force between them. As with magnetic attraction, as the temporal particles associated with the magnetic fields spin around, the bonds they have with the temporal particles of Earth's gravitational field are constantly broken. Thus, the planet's gravitational field never gets a good grasp on the levitating object's magnetic field, such that the energy released by the constant breaking of the bonds between the temporal particles of the magnetic field of the tabletop object and the magnetic field of the levitating object is enough to over-come the gravitational attraction between the levitating object and the Earth (*Figure 3.25*). Of course, in this case and the one above involving magnetic attraction, the gravity being referred to is essentially everyday gravity. Under extreme conditions, gravity may win. Also, the mass of an object can be too large for its magnetic field to overcome even everyday gravity; its mass and the strength of its magnetic field have to have the right balance. Of course, the strength of the magnetic field of the other object in the interaction is also an important factor.

The Electromagnetic Force and the Electric and Magnetic Forces Revisited

Recall that an electrically charged particle such as an electron will temporally polarize interior space, such that the field lines start on the charged particle and radiate outward. When the field line system is stationary, it is simply referred to as an electric, electrostatic, or static electric field. When the field

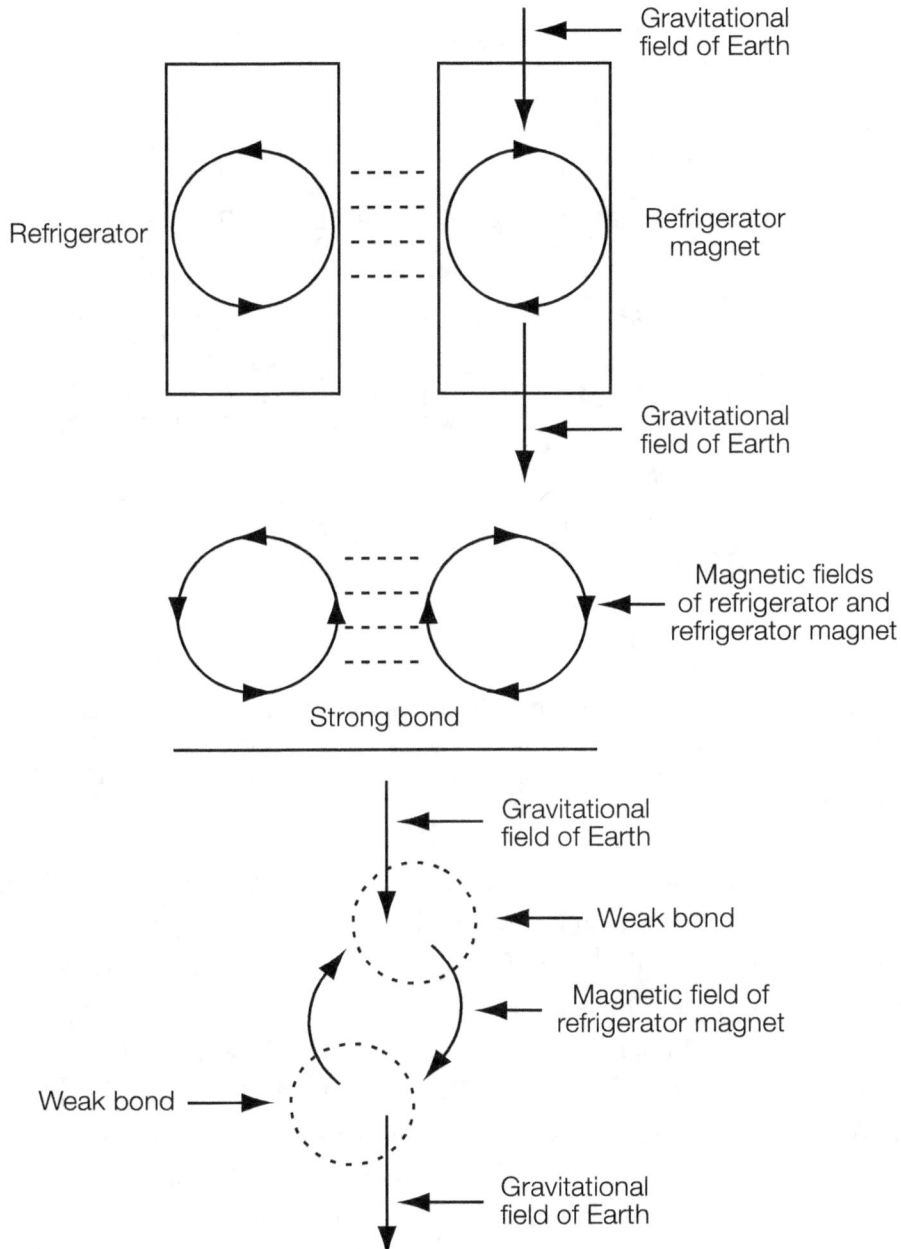

Figure 3.24 *Interactions between temporal particles involved in magnetic attraction and gravitation. The magnet is able to defy gravity because gravitational attraction is weakened by the cycling of the temporal particles in the magnetic field, such that the bonds between those temporal particles and those of Earth's gravitational field are constantly weakened/broken.*

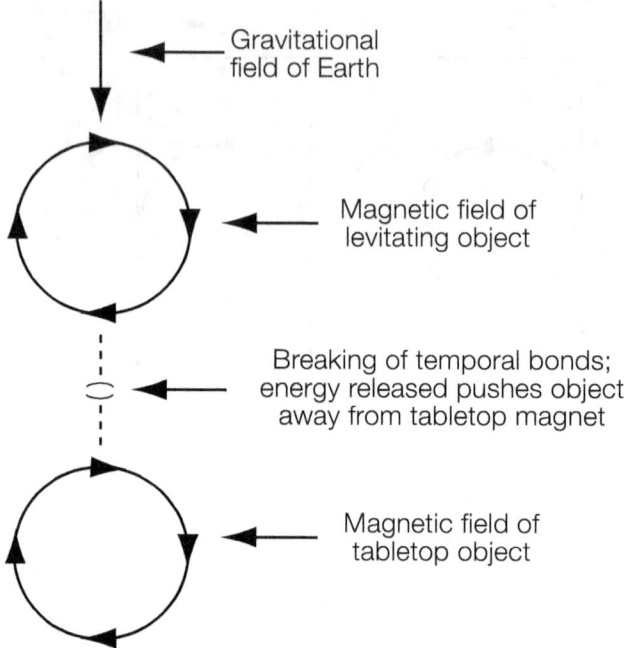

Figure 3.25 *Interactions between temporal particles involved in magnetic repulsion and gravitation. The energy released by the breaking of the bonds between the temporal particles of the tabletop and levitating objects' magnetic fields overcomes gravity's pull on the levitating object.*

line system moves, due to for instance the motion of its source particle, it develops a magnetic nature and is referred to as an electromagnetic field. This new field typically takes the form of waves that propagate away from the source particle. For example, consider a stationary electron and its static electric field. Recall also that polarized lines always occur in pairs. Every blue line has a yellow partner or vice versa. In TET, if the electron is jerked downward, an initial downward-pointing wave is created in the blue line immediately adjacent to the particle. This movement causes an equal, yet opposite reaction in its yellow partner, creating an upward-pointing yellow wave, which induces the formation of the next downward-pointing blue wave, which induces the formation of the next upward-pointing yellow wave, and so on. You can think of there being a tight junction between the blue and yellow lines that prods them into maintaining as straight a line as possible, almost like a seesaw. Thus, when one moves (that is, forms a wave pattern), the other does the

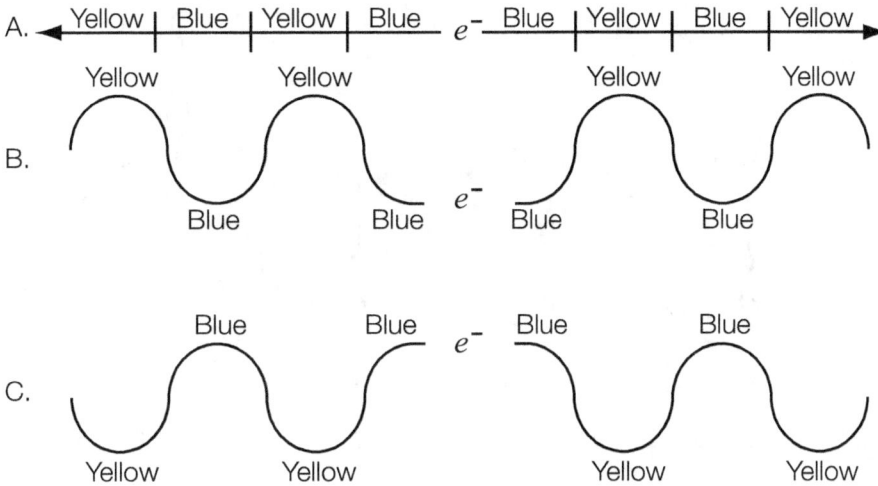

Figure 3.26 *Part A of figure represents a static electric field surrounding an electron. Parts B and C represent the transformation of the electric field into an electromagnetic field due to the jerking of the electron down (part B) and up (part C), although the magnetic component is not shown.*

same, but oppositely. The only difference is that the initial waves do not have the same look as the following waves, simply because they originate on the source particle, in this case an electron, rather than coming somewhere in the middle of a field line. Note that if the electron were jerked upward, the situation would be the same except in reverse (*Figure* 3.26).

As indicated above, blue elements attract and form bonds with t^+ particles, and yellow elements attract and form bonds with t^- particles. As the blue and yellow waves form, more t^+ and t^- particles move toward them from the top and bottom due to the increased energy, creating their magnetic fields (*Figure* 3.27). (Actually, the particles are coming in from all sides; it is just that the connection of the waves to each other makes the side influx less pronounced, except on the ends.) Note that the magnetic fields of the yellow waves are composed of t^- particles. Because magnetism largely involves interactions between temporal particles and because temporal particles do not recognize each other's spatiotemporal charge, forming bonds with each other regardless of those charges, a magnetic field composed of all t^- particles will work similarly

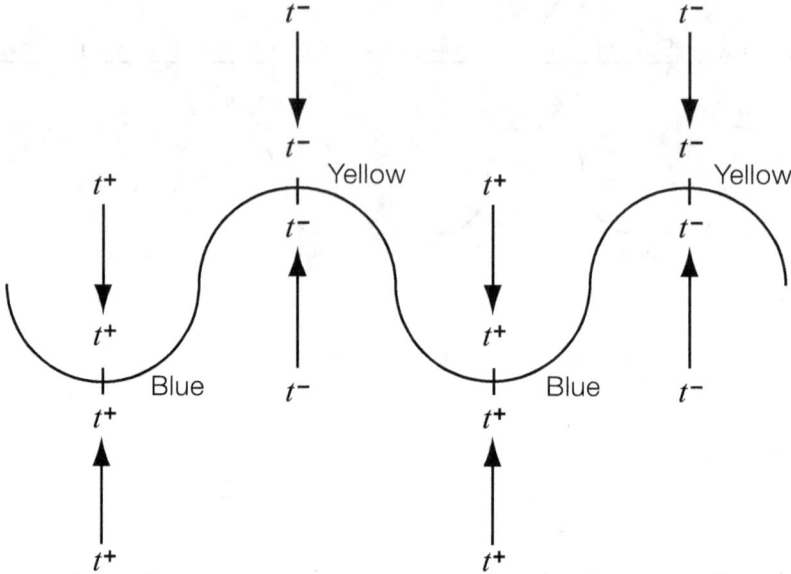

Figure 3.27 *Blue elements attract and form bonds with t⁺ particles, and yellow elements attract and form bonds with t⁻ particles. As the blue and yellow waves form, more t⁺ and t⁻ particles move toward them from the top and bottom due to the increased energy, creating their magnetic fields.*

to one composed of all t⁺ particles. The main difference of course is that a full magnetic effect involves the t⁺/t⁻ process, with the ultimate loss of t⁻ to exterior space.

With their motion energy, magnetic field energy, the energy associated with their blue and yellow lines—with blue representing positive energy and yellow representing negative energy—and the collective energy of the temporal particles associated with those lines, the waves behave like little packets of energy and can be thought of as particles. These particles are called particles of light, quanta of the electromagnetic field, electromagnetic radiation, or simply photons. They are the phenomena that allow us to see. As electric field lines occur in the blue-yellow format, so do the photons that arise from them. As temporal polarization is the same thing as electric charge, photons, in TET, are electrically charged particles, with one being negatively charged and the other positively charged. The two together, as a single pair, are called a neutral photon (*Figure 3.28*).

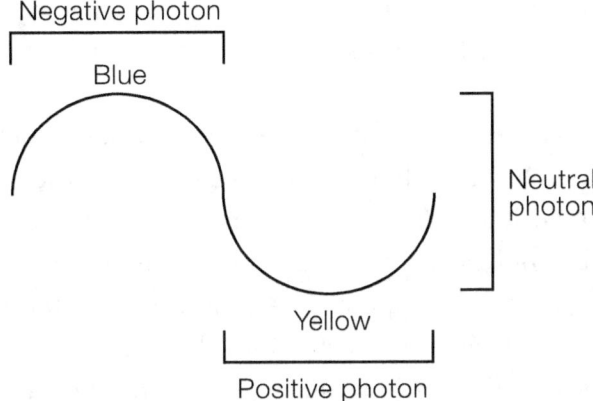

Figure 3.28 *In TET, a typical photon (an electrically neutral particle) is composed of a negative photon and a positive photon. The magnetic components of the photons are not shown in figure.*

Earlier in this chapter, it was stated that QM describes electric and magnetic phenomena as being brought about by electrically charged particles exchanging photons. (The photons being referred to were electrically neutral, because in QM there are no photons with electric charge.) In QM, photons are also capable of causing an impact force that can alter another particle's position and momentum due to the photons' particle-like nature, similar to one billiard ball hitting another. TET agrees with the concept of photons behaving like massive particles that can alter another particle's position and momentum, but it differs somewhat with the concept of photons causing the electric and magnetic forces; recall that there was no mention of them in TET's description of electric and magnetic phenomena.

In QM, because photons are thought to be responsible for both electric and magnetic phenomena, these two forces are often generally referred to as a single phenomenon called the electromagnetic force. Interestingly, QM considers photons to be responsible for the electric force even when the electric field is static. This may seem strange because only electric fields that are moving create photons. In QM, the photons that are thought to be exchanged in an electrostatic field are considered to be *virtual photons*—photons that are exchanged so quickly they cannot be detected. These undetectable photons are also thought to be exchanged in the magnetic fields brought about through the spins of particles like the electron, a magnetic field that exists even when the electron (the electron's body) is sitting still.

TET posits, however, that the reason these photons are undetectable is simply because they are not there. Photons are created when there are disturbances in an electric field line. When there are no such disturbances, there are no photons. In TET, only static electric field lines and temporal particles—the pure ingredients to make photons—exist around stationary electrically charged particles and cause an electric force between them. These ingredients are also responsible for causing a magnetic force between two stationary charged particles. In this case, the static field line component is simply the body of the charged particle itself, such as the blue circle representing the electron. Recall that this circle is simply a blue electric field line, except separated from its yellow partner and wrapped back on itself. As noted earlier, for magnetism in its fullest sense to occur, a particle must be able to process time in the t^+/t^- direction, which involves it having a blue (including partly blue) nature. (Only the positron is completely yellow; all of the other QM particles can experience full magnetism.) Thus, the body of the electron, in addition to the temporal particles associated with its spin, help bring about spin-related magnetism.

In TET, it is the presence of the ingredients to make photons that gives the impression that photons are present and being rapidly exchanged between two stationary particles interacting electrostatically or two such particles interacting magnetically through their spins only. Thus, referring to the particles as *virtual photons* is not so incorrect, because it is true that they are *virtually* there, because their ingredients are there. However, the concept that photons are being exchanged very quickly—too quickly to be detected—is not required in TET. All that is needed to bring about the electric and magnetic forces—indeed whether the source particles are moving or are stationary— are electric field lines (particularly blue lines) and temporal particles. That is, if these ingredients are in the form of real photons, that is fine; if not, that is fine too. The electric and magnetic forces will work well regardless of whether or not the ingredients creating them are in the form of real photons.

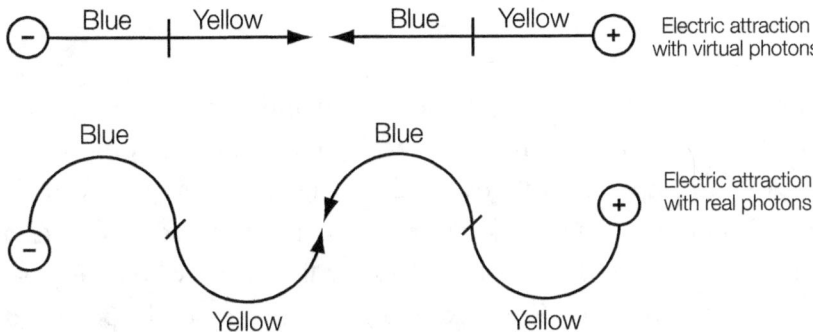

Figure 3.29 *In TET, the electric force works the same whether virtual or real photons are present.*

With regard to the electric force, real photons are created when electric field lines move up or down, as described above (*Figure 3.29*). The real photons at the heart of magnetism are the charged particles themselves as they move through space. The bodies of these particles are electric field lines; thus when they are stationary, they represent virtual photons, and when they move, real photons. (To help in understanding how an electrically charged particle can be regarded as a photon, recall that, in TET, there are negative and positive photons, in addition to neutral photons. An electron, for example, is simply a negative photon separated from its positive partner.) When charged particles move, small waves are created within their bodies (*Figure 3.30*). Note, however, that

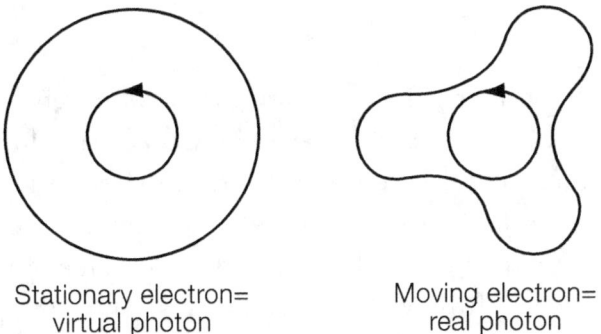

Stationary electron=   Moving electron=
virtual photon   real photon

Figure 3.30 *The real photons at the heart of magnetism are the charged particles themselves as they move through space. The bodies of these particles are electric field lines. When charged particles move, small waves are created within their bodies. Thus, when they are stationary, they represent virtual photons, and when they move, real photons.*

these waves are not further polarized; for example, the waves within a moving electron are all blue.

In QM, virtual photons (referring to neutral photons) are said to exist on energy borrowed from the vacuum. In TET, the ingredients of these photons—static field lines and temporal particles—indeed possess energy. However, energy need not be borrowed from the vacuum to create the static field lines because the lines are the vacuum. Recall that the field lines arise when areas of interior space—the vacuum of our direct experience—are temporally polarized. Thus, they are and remain just a part of that space. The energy of their existence is the same energy that exists within the fabric of interior space itself, just proportioned differently. What was once green energy is now blue-yellow energy, but the amount of energy remains the same (*Figure 3.31*). With regard to the temporal particles, however, energy is indeed borrowed from the vacuum. Note that the field lines come from interior space only, but temporal particles bounce back and forth between interior space and exterior space, so both of these vacuums should be considered.

Figure 3.31 *The simple polarization of an area of interior space does not change its energy level. With polarization, the energy is simply proportioned differently. What was once green energy is now blue-yellow energy, but the amount of energy remains the same.*

From the standpoint of interior space, ignoring for the moment its interactions with exterior space, the t^+ particles that are borrowed as part of the ingredients of the virtual photons never cease being a part of that vacuum even as they are borrowed, because interior space houses t^+ particles. That is, there is a constant presence of t^+ particles in interior space, and some of these particles serve as the ingredients of the virtual force carriers. Thus, the payback of t^+ energy is a non-issue from the standpoint of interior space. Being green, interior space also contains some t^- particles. As with the energy of the field lines, the t^+ and t^- energy in interior space can be thought of as being the same before and after polarization, just proportioned differently. Also, recall

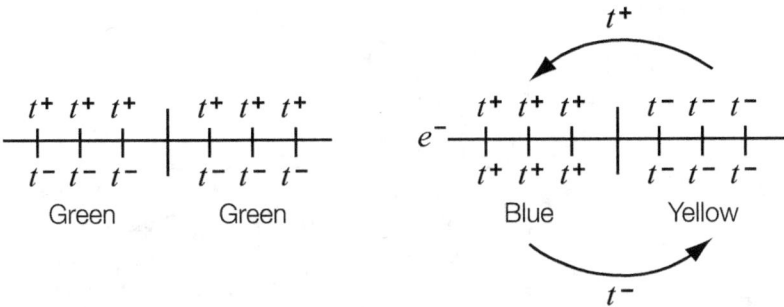

Figure 3.32 *The yellow component of a static electric field line obtains the t^- it needs from its blue sister. Ignoring interactions with exterior space, no extra energy needs to be added to the system. Note the segregation of t^+ and t^- in the electron image.*

that, upon polarization, any t^- needed in an electric field line will come from the processing of some of the t^+ particles in that field line. The main idea here is that no extra t^+ is added to the system. Up to this point, the energy of the virtual photons—both the field-line energy and the temporal-particle energy—is the same energy that exists within interior space itself. The energy remains a part of interior space even as it is borrowed to make the virtual photons, making payback, again, a non-issue (*Figure 3.32*).

From the standpoint of exterior space, however, energy is borrowed over time from that space and is quickly paid back. It is borrowed from exterior space through the t^-/t^+ process in TET and through an equivalent process in QM, which allows energy to be borrowed from the vacuum in conjunction with nothing more than time progression. From the perspective of TET, QM's borrowed energy is t^+ energy and QM's time is t^-/t^+. The borrowed energy is used to create the virtual photons. However, in QM, there is no specific mechanism by which the energy is paid back to the vacuum; it just is. In TET, the payback occurs through the t^+/t^- process. That is, the temporal energy borrowed from exterior space in the form of t^+ is quickly returned to that space in the form of t^-. Although both the t^+/t^- and t^-/t^+ respiration processes occur in an electric field line, ultimately the field line loses energy through the t^+/t^- process, or rather it gives back to exterior space the energy it took. Likewise, particles such as electrons borrow t^+ energy and quickly pay it back in the form of t^- (*Figure 3.33*). In a sense, interior space can be thought of as the

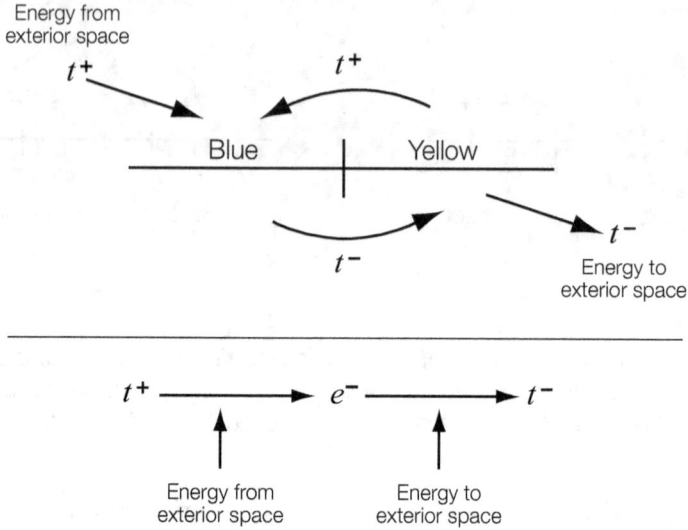

Figure 3.33 *Shown are a virtual photon and an electron borrowing energy from exterior space as t⁺ and paying it back as t⁻.*

"middleman" between exterior space and essentially anything that processes time in the t^+/t^- direction. Exterior space can be considered to give t^+ to interior space, and interior space gives it to, for instance, an electric field line or an electron, which then returns the energy to interior space as t^-, which exterior space absorbs. Ultimately, the energy is being borrowed from exterior space and returned to that space; interior space just passes it along.

Interestingly, the t^-/t^+ process in TET (and in QM, another equivalent phenomenon when QM is combined with Special Relativity, chapter 6) is used to create real photons, as well. Earlier in this chapter, it was stated that force carriers come from the vacuum, which can be regarded as interior space plus time as a single, unified field. To help illustrate the rise of the photons from the vacuum, consider a single line of interior space. This line is colored green and has an even distribution of t^+ and t^- particles across it. When an electrically charged particle such as an electron is placed in the vicinity, it temporally polarizes the green line into blue and yellow lines, with the t^+ moving to the blue line and the t^- moving to the yellow line. This scenario represents a static electric field line, which is the same as saying a virtual photon. Note that the change from green to blue and yellow and the redistribution of t^+ and t^- energy has not altered the overall energy of the line—that is, the blue-yellow line has the same energy as the green line, just proportioned differently. For this reason, this electric field line/virtual photon is indistinguishable from the vacuum itself. It is almost as if there is nothing there. Although another charged particle would sense the presence of the field line/virtual photon (due to the polarization), beyond this, it is no different from what is generally thought of in everyday language as "empty space."

For a real photon to be created, energy must be applied to the electric field line/virtual photon, perhaps by the jerking of the electron. As described previously, this leads to the creation of an electromagnetic field line, which is the same as saying real photons, consisting of waves in the electric field line plus magnetic fields associated with those waves. However, ignoring the magnetic fields for a moment, the simple creation of waves in the electric field line is insufficient for creating the real photon. Imagine for a moment that the green and blue-yellow lines above were shaped like waves. There is nothing about the wave pattern in and of itself that makes the blue-yellow line stand out from the vacuum from the standpoint of energy. Even in that formation, it is indistinguishable from the vacuum. What creates a real photon is the influx of t^+ particles in response to the energy that created the waves in the first place, the jerking of the electron in the case of the electric field line/virtual photon. As noted earlier, when temporal particles congregate, they compress the space between them. The blue-yellow line is space. Thus, as the t^+ particles surround the line (which is now in a wave pattern), they compress it, which concentrates the energy associated with it. The concentration of the energy associated with the blue-yellow waves makes the waves stand out from the vacuum, helping the photon to become real (*Figure* 3.34). The energy of the magnetic fields within the waves also comes from the t^-/t^+ process and further helps the photon to become real.

With the newly supplied t^+ energy, responding to the energy of the jerking of the electron, the waves have become something discernible, something that stands out from empty space. It is for these reasons in TET that, like the virtual photons, real photons depend on the t^-/t^+ process, as this ultimately is the

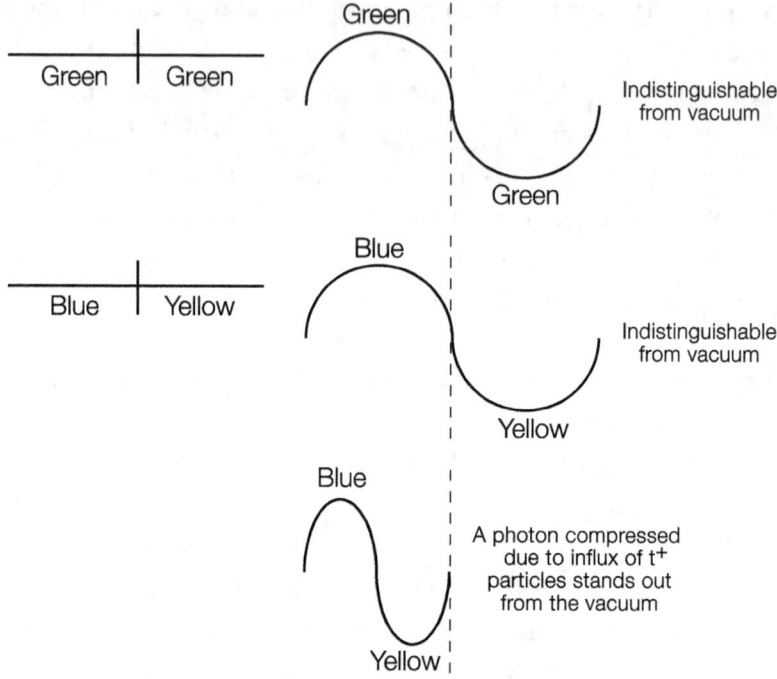

Figure 3.34 *Ignoring the magnetic fields for a moment, the simple creation of waves in an electric field line is insufficient for creating a real photon. The waves must be compressed for the photon to stand out from the vacuum.*

source of the (t^+) energy needed to create them. Matter and anti-matter particles, such as electrons and positrons, by their very nature (as compressed objects) stand out from the vacuum, but as they move through space becoming real photons, they too are compressed by an influx of t^+ particles, making them stand out from the vacuum that much more.

Photons—real and virtual—indeed create the electric and magnetic forces and thus are a link between the two. As noted earlier, because of this, they are usually referred to collectively as the electromagnetic force. (Although this term will often be used in later discussions to refer to the two forces, sometimes it will be better to refer to them separately.) Whereas electric field lines and temporal particles create the electromagnetic force, weak field lines and temporal particles create the weak force, and strong field lines and temporal particles create the strong force. As with the electromagnetic force, the concept of force carriers existing for such brief amounts of time that they cannot be detected is not needed in TET to describe the weak and strong forces. It is really the ingredients of their force carriers that create the forces. Before the strong and weak forces are discussed in more detail, TET's description of the creation and structure of elementary QM particles as well as other important concepts will be introduced. These discussions will aid in TET's description of the weak and strong forces.

TET and the Creation and Structure of Elementary QM Particles

Typically, elementary QM particles are created through an electromagnetic field. The weak and the strong fields, along with their associated magnetic fields, may also give rise to these particles; this discussion, however, focuses on the electromagnetic field. From the perspective of TET, in the early universe, in which there were originally no QM particles, both interior and exterior space fluctuated and expanded wildly, and this instability caused random polarizations, random electromagnetic fields, to arise in interior space. It was from these initial electromagnetic fields that all QM particles were created. For simplicity, the discussion on creation in this chapter focuses only on the photon, electron, and positron. However, all the elementary QM particles are discussed with regard to their structure. The topic of particle creation is revisited in more detail in chapters 8 and 9.

All QM particles in TET are characterized by three components: (1) a frame, (2) a mass field or simply its mass, and (3) a spin field or simply its spin (*Figure* 3.35). The frame of a QM particle is a well-defined area of space with energy that is usually higher than the energy of the background, that is, of interior space. The frame is usually, but not always, temporally polarized also. In the case of the electron, it is the blue electric field line that is wrapped back on itself. The mass of a QM particle is the external volume of temporal particles that are responding specifically to the energy of the frame, with some of those particles directly bound to the frame. The spin is a region of moving space within the QM particle. Associated with this space is another volume of temporal particles (an internal volume) that is responding to the energy of the spin and also to the energy of the frame on that side (the internal

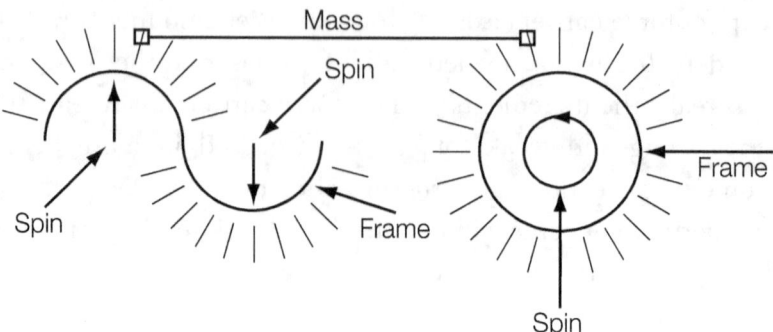

Figure 3.35 *All QM particles in TET are characterized by three components: (1) a frame, (2) a mass field or simply its mass, and (3) a spin field or simply its spin.*

side). The volume of temporal particles responding to the spin creates an internal magnetic field within the QM particle, as described previously.

The frame of a neutral photon comprises both its blue and yellow electric field lines. As the waves constituting a real photon form, they develop an "inside" and "outside." Although each wave's magnetic field technically exists all around it, forming as a result of nothing more than the movement of space, there is a definite cupping or enclosing of part of that magnetic field by each wave. The enclosed portion of the magnetic field, or rather the moving space that the magnetic field is responding to, is the spin of the wave—with the spin of both waves constituting the spin of the photon. The temporal particles associated with the waves on their more "pointy" (external) sides, those responding directly to the energy of the frame, constitute the photon's mass. (Note that it is understood that [neutral] photons have mass; they simply do not have rest mass, as there is no way to make them come to a rest.) Although technically not present due to the movement of space, the temporal particles associated with the more internal sides of the waves/frames can, for general accounting purposes, be counted as part of the internal magnetic field of the photon. And although strictly speaking, spin is the area of moving space within a particle, for ease and simplicity, the term can be used to refer to that moving space plus the magnetic field because they go hand in hand.

When enough energy is applied to an electromagnetic field, negative and positive photons can separate, with their field lines closing to form circles.

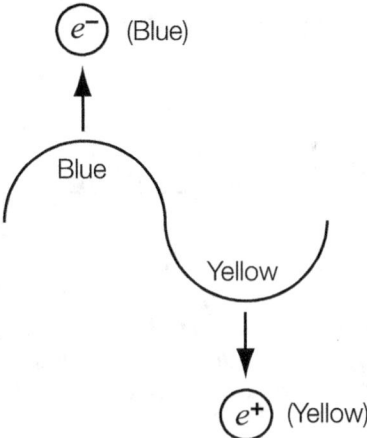

Figure 3.36 *When enough energy is applied to an electromagnetic field, negative and positive photons can separate, with their field lines closing to form circles. These circles represent the electron and the positron.*

These circles represent the electron and the positron (*Figure* 3.36). Essentially, the electron is a negative photon, and the positron is a positive photon. As stated earlier, the electron is called a matter particle, and the positron is an anti-matter particle. Matter and anti-matter are created from light in equal amounts. In general, a matter particle and its anti-matter partner have opposite electric charges. If the electron and positron were to meet again, they would simply re-form into light, although the half waves would move in opposite directions (*Figure* 3.37). From a temporal perspective, the electron looks as though it has positive spatial energy (its "blueness") and is moving forward in time (t^+ to t^-), whereas the positron looks as though it has negative spatial energy (its "yellowness") and is moving backward in time (t^- to t^+).

Particle Frames

(Note that this section deals only with elementary matter and anti-matter particles. The characteristics of the frames of the force carriers are discussed in later sections.) The frames of elementary matter and anti-matter particles house their temporal polarization (and thus electrical nature) and some of their overall energy. For an electron, it is simply a blue electric field line

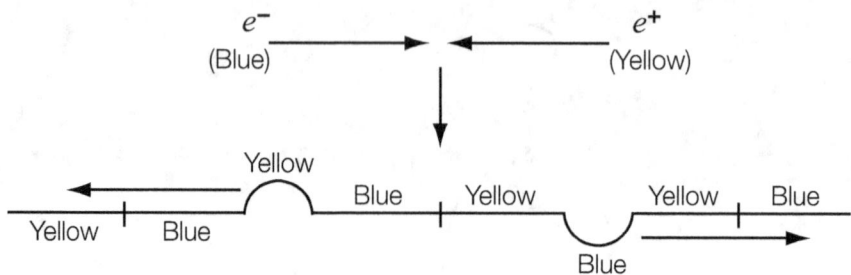

Figure 3.37 *If an electron and positron were to meet again, they would simply re-form into light, although the half waves would move in opposite directions. The half waves oscillate as they travel and alternate being blue and yellow.*

wrapped back on itself. It is helpful to think of the frame of elementary matter and anti-matter particles as being composed of smaller segments. Each of these smaller segments represents one third of electric charge, negative or positive, or the lack thereof. The frames of elementary matter and anti-matter particles can contain only three, four, or five of these segments. With six, you get two separate frames of three segments each. Also, up to three of these segments can be polarized, electrically charged, in a given elementary matter or anti-matter particle. Any other segments would be true green.

Table 3.5 lists several of the elementary matter and anti-matter particles along with their characteristic segments. An electron is represented by three blue segments, a muon by three blue and one green, and a tauon by three blue and two green segments. A positron is represented by three yellow segments, an anti-muon by three yellow and one green, and an anti-tauon by three yellow and two green segments. Thus, each has one unit of electric charge, which is negative for the electron, muon, and tauon, and positive for the positron, anti-muon, and anti-tauon. True-green segments mixed with fully polarized segments leads to instability within frames. Using a weather analogy, a green segment mixed with, for example, blue segments is like a cold front and a warm front interacting to create a thunderstorm. Muons, anti-muons, tauons, and anti-tauons are highly unstable; the energy associated with this instability attracts more t^+ particles, making them appear more massive. The more green segments they have the more massive they appear to be.

Table 3.5 *TET Description of Electron, Muon, and Tauon, and Their Anti-Matter Particles*

Generation		Matter			Anti-Matter			
		Particle	TET Symbol	TET Structure	Electric Charge	Particle	TET Symbol	TET Structure	Electric Charge
I	e^-	bbb	(triangle of b, b, b)	−1	e^+	yyy	(triangle of y, y, y)	+1
II	μ^-	$bbbg$	(diamond of b, b, b, g)	−1	μ^+	$yyyg$	(diamond of y, y, y, g)	+1
III	τ^-	$bbbgg$	(pentagon of b, b, b, g, g)	−1	τ^+	$yyygg$	(pentagon of y, y, y, g, g)	+1

Earlier it was stated that in general, matter and anti-matter particles have opposite electric charges. This is also true for neutrinos, even though they are typically referred to as electrically neutral. The reason for this is that, in TET, neutrality is not necessarily absolute neutrality. A segment that is green may indeed be truly green, or it may favor the blue side of green or the yellow side of green. The green segments in muons, anti-muons, tauons, and anti-tauons are true-green segments. Recall that each has three fully polarized segments, so they cannot accept any more polarized segments, even ones that are slightly polarized. However, particles with zero, one, or two fully polarized segments may contain slightly polarized (bluish-green or yellowish-green) segment(s).

For example, an electron neutrino has no fully polarized segments, but has three bluish-green segments. Thus, by being green, it is technically electrically neutral, but its slight favoring of blue gives it a slight electrical nature. The anti–electron neutrino has three yellowish-green segments. The muon neutrino has three bluish-green segments and one true-green segment, because again, there can be no more than three fully or slightly polarized segments within a single particle. The anti–muon neutrino has three yellowish-green segments and one true-green segment. The tauon neutrino has three bluish-green segments and two true-green segments, and the anti–tauon neutrino has three yellowish-green segments and two true-green segments (*Table 3.6*). The

Table 3.6 *TET Description of Neutrinos and Anti-Neutrinos*

		Matter			Anti-Matter		
	Particle	TET Symbol	TET Structure	Electric Charge	Particle	TET Symbol	TET Structure	Electric Charge
I	ν_e	$\dfrac{bbb}{yyy}$	b/y b/y b/y	0	$\bar{\nu}_e$	$\dfrac{yyy}{bbb}$	y/b y/b y/b	0
II	ν_μ	$\dfrac{bbb}{yyy}g$	b/y b/y b/y g	0	$\bar{\nu}_\mu$	$\dfrac{yyy}{bbb}g$	y/b y/b y/b g	0
III	ν_τ	$\dfrac{bbb}{yyy}gg$	b/y b/y b/y g g	0	$\bar{\nu}_\tau$	$\dfrac{yyy}{bbb}gg$	y/b y/b y/b g g	0

(Generation)

presence of true-green segments in the muon and tauon types of neutrinos does not add instability because there are no fully polarized segments in those particles. Using the weather analogy again, it is as if to say there is not as extreme a "temperature" difference between bluish-green segments, for example, and true-green segments as there is between blue segments and true-green segments.

TET considers some neutrinos to exist that are true-green all over. Such neutrinos are called sterile. Like the other particles, they can have three, four, or five segments. There are three types of sterile neutrinos: sterile electron neutrinos, sterile muon neutrinos, and sterile tauon neutrinos (*Table 3.7*). Because they are composed of only true-green segments, they are their own anti-matter particles. Thus, there really is no such thing as a "sterile anti–electron neutrino," just a sterile electron neutrino. Because they are not a part of QM, sterile neutrinos are considered QM-type particles.

Another important point that should be mentioned is that while three segments can be temporally polarized within a given elementary particle, there cannot be polarized segments that are starkly opposite. That is, an elementary particle cannot have both blue and yellow segments in its frame. Blue and yellow are far too opposite for a particle to contain both of them and be stable. Likewise, a particle cannot have both bluish-green and yellowish-green segments in its frame. Bluish green and yellowish green are far too opposite. Interestingly, a particle

Table 3.7 *Sterile Neutrinos in TET*

	Particle	TET Symbol	TET Structure	Electric Charge	Short-hand
I	Sterile electron neutrino	ggg	$\underset{g\diagdown\ \diagup g}{\overset{g}{\frown}}$	0	ν_e^s
II	Sterile muon neutrino	$gggg$	$\underset{g\diagdown\ \diagup g}{\overset{g\frown g}{}}$	0	ν_μ^s
III	Sterile tauon neutrino	$ggggg$	$\underset{g}{\overset{g\frown g}{g(\ \ )g}}$	0	ν_τ^s

Generation

can have blue and yellowish-green segments or yellow and bluish-green segments. Blue and yellowish green do not clash, nor do yellow and bluish green. Frames with blue and bluish-green segments and those with yellow and yellowish-green segments are unlikely to occur.

This idea leads into the next set of particles—the quarks. These particles are composed of a mixture of fully polarized segments and slightly polarized segments. For example, an up quark is composed of two yellow segments and one bluish-green segment—it thus has an electric charge of +2/3. An anti–up quark is composed of two blue segments and a yellowish-green segment—it thus has an electric charge of –2/3. A down quark is composed of one blue segment and two yellowish-green segments—it has an electric charge of –1/3. An anti–down quark is composed of one yellow segment and two bluish-green segments—it has an electric charge of +1/3. There are six matter quarks and six anti-matter quarks *Table 3.8*.

Mass

As QM particles form, be they force carriers or matter or anti-matter particles, the temporal particles associated with the external side of their frames become their mass—these are temporal particles responding specifically to the energy of the frames, with some of these particles directly bound to the frames. However, it is important to note that it is the motion energy of the

Table 3.8 *TET Description of Quarks and Anti-Quarks*

Generation		Matter			Anti-Matter		
	Particle	TET Symbol	TET Structure	Electric Charge	Particle	TET Symbol	TET Structure	Electric Charge
I	u $yy\frac{b}{y}$	TET Structure	+2/3	$\bar{u}$ $bb\frac{y}{b}$	TET Structure	−2/3
I	d $b\frac{yy}{bb}$	TET Structure	−1/3	$\bar{d}$ $y\frac{bb}{yy}$	TET Structure	+1/3
II	c $yy\frac{b}{y}g$	TET Structure	+2/3	$\bar{c}$ $bb\frac{y}{b}g$	TET Structure	−2/3
II	s $b\frac{yy}{bb}g$	TET Structure	−1/3	$\bar{s}$ $y\frac{bb}{yy}g$	TET Structure	+1/3
III	t $yy\frac{b}{y}gg$	TET Structure	+2/3	$\bar{t}$ $bb\frac{y}{b}gg$	TET Structure	−2/3
III	b $b\frac{yy}{bb}gg$	TET Structure	−1/3	$\bar{b}$ $y\frac{bb}{yy}gg$	TET Structure	+1/3

temporal particles that ultimately is mass, not the temporal particles themselves. For example, imagine a fairly wide jar with a small ball in the bottom. Next imagine a disk that fits into the jar and falls to the bottom on top of the ball. The disk is of course not level because part of it is raised up by the ball. Now imagine the ball revolving around the bottom of the jar with incredible speed. Under this condition, the disk is able to be level. The ball is moving so fast that it is as if the disk were sitting on a large solid, level mass at the bottom of the jar. This is analogous to the mass generated by the motion of temporal particles.

Also, earlier it was stated that from a temporal perspective, electrons look as though they have positive energy and are moving forward in time (t^+/t^-) and that positrons look as though they have negative energy and are moving backward in time (t^-/t^+). This energy has to do with the energy of their frames—

with blue representing positive energy and yellow representing negative energy—but this concept also extends into mass. That is, in the strictest of senses, the electron has positive mass, and the positron has negative mass. It is important to understand the link between mass and energy in this sense: The electron's blue frame takes t^+ temporal energy and converts it first to t^0 and then to t^-. Thus, the electron looks as though it has positive mass, positive energy (which it is continuously extracting from its mass field), and that it is moving forward in time. The positron's yellow frame takes t^- temporal energy and converts it first to t^0 then to t^+. Thus, the positron looks as though it has negative mass, negative energy (which it is continuously extracting from its mass field), and that it is moving backward in time.

There are a few caveats to the idea of negative mass: First, temporal particles interact with each other in the same manner regardless of their spatiotemporal charge. The charge on a temporal particle only relates to how that particle will interact with polarized, non-polarized, or slightly polarized space, be that space in the form of a particle frame or whole vacuum, such as exterior space. Although there are some differences with regard to gravity when a particle has negative mass, those differences are not major factors—they will be explored in chapter 5. Also, quarks, muons, tauons, neutrinos, and all of their anti-matter counterparts have mixed mass fields, owing to their green segments. Probably the greatest examples of mixed mass fields are those of atoms, which are basically complexes of electrons and quarks. Thus, our experience with mass is mostly with mixed fields, making it seem as though there is really only one type of mass (mixed), which in the practical sense is either there (positive) or not (zero). Moreover, it is the motion energy of temporal particles that ultimately is mass. Those particles could have a positive or negative spatiotemporal charge; it does not much matter. Therefore, this is another practical sense in which all mass can be regarded as positive, because this motion energy is either there (positive) or not (zero). The positron can be regarded as having negative mass or positive mass. Taking the charge of the temporal particles into account, it has negative mass. Taking just the energy of the temporal particles in its mass field into account, it has positive mass. Moreover, all mass, except for that of the positron, can be viewed as t^+, because the t⁻ used by particles is in most cases produced by the blue elements in those particles through the t⁺ that continuously flows into them.

Lastly, mass is for the most part only discernible when it is associated with a fully or partially polarized segment. Green segments blend in with interior space and thus so does the mass associated with them, even when the segment is compressed. All sterile neutrinos, therefore, would appear massless. All other neutrinos appear to have only a slight amount of mass, corresponding to their slight polarization. Electrons and positrons, of course, appear very massive, owing to their fully polarized segments. Also, the electron and positron have the same amount of mass; it is just positive in the case of the electron and negative in the case of the positron, although both can be viewed as having positive mass as described above.

Technically, the muon, tauon, and their anti-matter sisters have the same mass as the electron and positron, as they contain three fully polarized segments in their frames. The green segments of the muon, tauon, and their anti-matter particles are technically massless. These particles appear very massive in nature, however, because of the instability their green segments bring to their frames. The energy of the instability attracts more t⁺ particles to them, making them seem more massive than they are. Along this same line, charm, anti-charm, top, and anti-top quarks appear more massive than up quarks. And strange, anti-strange, bottom, and anti-bottom quarks appear more massive than down quarks. The muon, anti-muon, tauon, and anti-tauon neutrinos would likely be measured to have the same mass as the electron neutrino and the anti-electron neutrino, because their green segments do not add instability to their frames unlike they do to other particles.

Spin

As QM particles form, be they force carriers or matter and anti-matter particles, they develop an internal area of moving space. This space along with its associated temporal particles is a particle's spin. Although the spins of photons simply point up and down, the spins of some particles, such as electrons and positrons, do actually spin in TET. As electrons and positrons are created from photons, the spins of the photons transform into the spins of those particles. Recall the previous ideas that:

1. The electron is a little magnet in and of itself—the fundamental unit of magnetism in a bar magnet;

2. Every electron has inside of it a loop of spinning space;

3. Like a loop of electrical current, the spin of an electron develops a magnetic field around it.

Of course, the same is true for the positron. This little loop of "spatial current" develops as a result of the wave formation within the electric field and the direction the waves are traveling as the electron and positron are formed. Unfortunately, the TET description of spin formation in particles like electrons and positrons has a flaw. However, first, to determine spin, you have to think of there being vertical arrows within the undersides of the photon waves, pointing toward their peaks. Next, think of there being horizontal arrows starting at the heads of the previous arrows and pointing in the direction the waves are traveling. Now, to determine the spin of the electron and positron, consider the two waves to separate as usual, with the field lines closing to form the electron and positron's frames, and consider the arrows to be pointing

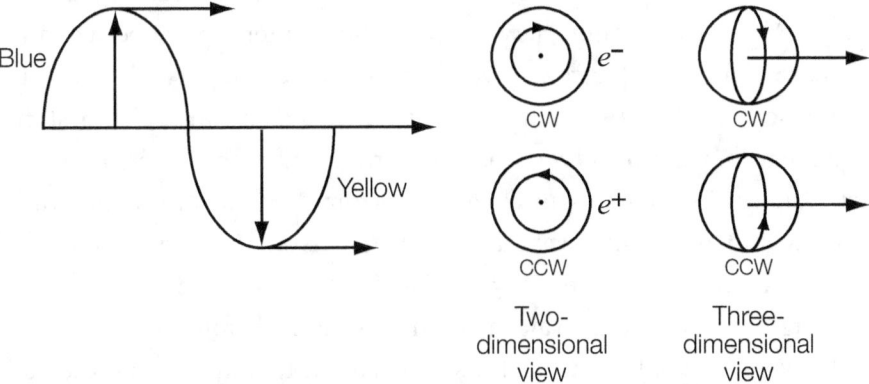

Figure 3.38 *In TET, the spin of photons gives rise to the spin of matter and anti-matter particles. TET does not place a particle's frame and spin in their three-dimensional cross pattern. That step must be done manually. Note that the spins in earlier figures were reckoned from the two-dimensional view.*

in the direction (clockwise or counterclockwise) of each particle's spin (*Figure* 3.38). Which one will be spinning clockwise and which one will be spinning counterclockwise depends on a number of factors such as whether the first wave of the electromagnetic field was up or down and whether the wave was emanating from a negatively or positively charged particle.

The flaw mentioned earlier is that this description makes it seem as though spin is in the same plane as a particle frame and that the particles are almost rolling in the direction of travel. The method gives a convenient, two-dimensional view of spin. Each spin and frame should actually be thought of as forming a cross pattern with each other. Also, in three-dimensional space, the spin would be the clockwise or counterclockwise movement as seen by an observer standing in the path of the particle and looking at it as it traveled toward him. Thus the flaw is that, although you can use TET to determine the spins of particles, you have to place that information into three-dimensional space yourself. TET does not do that automatically. That is, it does not auto-matically form the three-dimensional cross pattern between the spin and particle frame or properly orient the spin in its direction of travel. Those steps have to be performed manually.

Note that the linear movement of a particle and, as a consequence, the clock-wise/counterclockwise direction of its spin can appear differently to different observers. With everything viewed from the reader's perspective, imagine from left to right, an electron traveling toward a stationary observer, then a moving observer who is traveling away from the electron and the stationary observer. The moving observer would determine that the electron, as well as the "stationary" observer, is moving away from him, whereas the stationary observer would determine that the same electron is moving toward them both. The stationary observer may say the electron is spinning clockwise, but the moving observer would say the same electron is spinning counterclock-wise—he would consider the clockwise movement to only be what the counter-clockwise movement looked like from behind as the particle moved away from him. There is technically no "correct" answer, as all motion (and non-motion), even the reader's state, is relative. An observer standing in the path of a moving particle—who is stationary compared to the particle—is simply a good reference point. Note also that spin is below the level of friction—which occurs at or above the level of particle frames—and a body in motion remains in motion unless a force is applied to stop it.

As described above, the spinning space develops a magnetic nature due to the response of the temporal particles. The electron and positron, each with a spinning loop of space within it, takes on magnetic north and south poles. And again, the magnetic nature of bar magnets comes from the spins of the electrons within them. In some matter and anti-matter particles, the spin moves not in circles but in random directions. This would not produce a definite magnetic north or south, but the spin can still be considered a magnetic field. It is just that such a magnetic field is actually composed of many little magnetic fields moving in random directions. Lastly, the spin of QM particles appears to move faster than the speed of light. In TET, this is because the spin is ultimately measured via the temporal particles associated with it, and such particles do move faster than the speed of light.

Spatial and Temporal Energy and the Universal Clock

The energy of particle frames and mass and spin fields reduces to spatial energy, temporal energy, or some combination of the two. Electrons look as though they have positive spatial energy, owing to the "blue energy" of their frames, whereas positrons look as though they have negative spatial energy, owing to the "yellow energy" of their frames. The mass and spin fields of electrons are composed of t^+ energy, and the mass and spin fields of positrons are composed of t^- energy. There is also spatial energy associated with the spin fields, but in the case of spin, spatial energy refers mostly to the motion energy of interior space, which can either be present (positive) or not (zero).

From a temporal perspective, as opposed to a motion perspective, interior space, being green in the TET color scheme, possesses equal amounts of positive energy and negative energy. However, because it houses t^+ particles, and largely excludes t^- particles (from the standpoint of storage), it can be regarded as a "positive energy vacuum." As noted earlier, exterior space is yellow all over, and it houses t^- particles. Because of this, it looks as though it has negative energy and can be regarded as a "negative energy vacuum." Interestingly, QM also posits a negative energy vacuum. In QM, this vacuum is used to explain why electrons do not have negative energy. That is, in QM, it is thought that electrons can radiate so much of their energy away that they themselves eventually attain negative energy. However, the presence of the negative energy vacuum in that theory blocks electrons from becoming negative. Essentially, QM's negative energy vacuum hoards all the negative energy for itself, leaving none for the electrons. QM's concept of an electron taking on negative energy is equivalent in TET to an electron transforming all of the t^+ particles associated with it to t^-. If the t^- particles were never absorbed by exterior space, the electron would have negative energy, in the form of negative mass (ignoring for the moment the TET idea that electrons repel t^- particles). However, exterior space quickly absorbs the t^- particles that the electron produces. Thus, TET's negative energy vacuum does the same thing as QM's negative energy vacuum, which is hoard negative energy. As discussed earlier, particles that require t^- energy have some ways to get around exterior space's hoarding of it. However, the main point is that, despite some minor differences, the reason electrons do not have negative energy in QM is the same reason they do not have negative energy in TET, because of the presence of a negative energy vacuum.

Note that some versions of QM use an ad hoc process call renormalization to, at least mathematically, make the negative energy vacuum disappear. Despite its important function in QM, this vacuum is not considered to be something real or physical. Through the renormalization process, the every-day vacuum is considered to have just enough positive energy to cancel the effects of the negative energy. Because the negative energy is essentially still there, electrons still do not fall into the negative energy range, but there is a positive layer of energy over the negative energy vacuum hiding it from view. What the renormalization process is doing from the TET perspective is looking at exterior space only though the filter of interior space, the positive energy vacuum. Thus, the negative energy vacuum is still there, but it is buried under a layer of positive energy. Renormalization of the negative energy vacuum is thus a fairly benign procedure from the TET perspective. No harm is really done because, as TET suggests, there is indeed a positive energy vacuum (interior space) lying "on top of" the negative energy vacuum (exterior space). However, the negative energy vacuum should not be regarded as something strange in QM that needs to be hidden from view mathematically. It is a very important part of the overall universe.

Another important feature of the negative energy vacuum is that it behaves like one, very large positron. Like this particle, it is positively charged and moves backward in time. In TET, every matter, anti-matter, and force carrier particle is its own clock. All of these particles exist in interior space; thus, this space has many clocks, all with their own time (that happens to vary with the particles' position within a gravitational field or their motion through space, as discussed in chapters 5 and 6). Interior space itself moves equally forward and backward in time and, therefore, is timeless. In brief, there is no universal clock in interior space. However, because exterior space behaves as if it were one particle, it does have or, rather, is itself a single clock. The clock that is exterior space in TET is likely the universal clock, the time parameter, of QM. It is likely that mysterious thing that ticks away regularly "somewhere out there," against which QM events are measured. The single clock that is all of exterior space, in a sense, extends over all of interior space. One of the reasons space and time seem so separate in QM is probably because QM's clock ultimately resides in the negative energy vacuum, while we live in the positive energy vacuum. The negative energy vacuum is QM's clock/time, and the positive energy vacuum is QM's space. And they are indeed separate and distinct systems. Another reason space and time seem separate in QM is likely because QM, for the most part, ignores the negative energy vacuum, which in turn is probably the reason the origin of its time parameter is unknown.

Although the time of exterior space technically runs counter to our time, it can be used as a forward-running clock. For example, if I had a clock whose hands rotated backward, I could still use it to tell forward-running time. Eleven o'clock would just be one o'clock; ten o'clock would be two o'clock, and so on. Of course, the numbers themselves could also run counter to each other between the two clocks. That is, on the backward-running clock, there could be a 1 where there is an 11 on the forward-running clock, and so on. Either way, a clock whose hands rotated backward would be just as useful as a clock whose hands rotated forward. Note that thermodynamic events, such as the melting of an ice cube, occur the way they normally occur regardless of whether the clock's hands rotate backward or forward. That is, it is not that we, with our t^+/t^- time, see the ice melt and exterior space, with its t^-/t^+ time, sees the ice refreeze. Thermodynamic events, such as the melting of ice, are largely, but not fully, independent of t^+/t^- and t^-/t^+ time.

Important Relationship Between Particle Frames and Field Lines

An important relationship between the frames of electrically charged particles and the field lines emanating from them is that the initial primary field lines mimic the frames of the particles. For instance, it was stated earlier that negatively charged particles start the polarization process with blue lines, whereas positively charged particles start the polarization process with yellow lines. The reason for this is that these initial polarized lines are mimicking their source particles. They become a reflection of the particles themselves, just in the form of a line. Moreover, if you think of an electron as being composed of three blue segments of one third electric charge, the initial polarized lines surrounding the electron can also be thought of this way. Any polarized line, initial or otherwise, will of course have a sister line, and the alternating blue and yellow lines will continue throughout interior space (*Figure 3.39*).

The same occurs in the case of quarks, which produce not electric field lines but strong field lines. Thus, the initial polarized lines emanating from, for

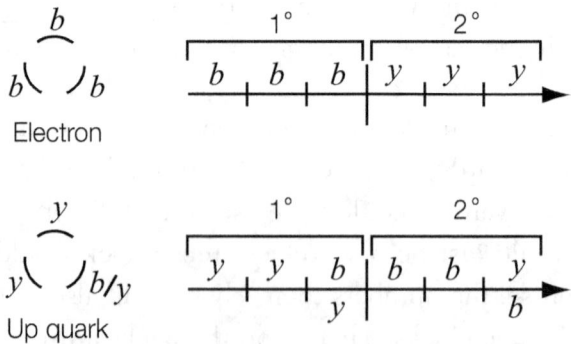

Figure 3.39 *The initial primary field lines emanating from charged particles mimic the frames of the particles, but in line form.*

example, an up quark, like the quark, will be composed of two yellow segments and one bluish-green segment (*Figure* 3.39). Note that when describing a mixed field line, particularly a line of force as opposed to a particle frame, it is helpful to note the fully polarized segment(s) first (in this case, the yellow segments), and the slightly polarized segment(s) after (in this case, the bluish-green segment) in the direction of the field line.

The field lines emanating from electrically charged leptons, quarks, as well as neutrinos do not form until such particles are fully formed. For example, electric field lines do not emanate from negative and positive photons until they have separated and the blue and yellow frames have wrapped back on themselves. This is because even in circle formation a blue or yellow field line is trying to create its sister, but in circle formation, it cannot do this directly. Thus, it projects itself outward—hence the mimicking aspect of the initial field lines—and creates its partner that way, although in all directions. As a straight line or open wave, the field line can induce the formation of its partner directly and thus has no need to project additional field lines from itself (*Figure* 3.40).

When particles bond with each other, they act as a single particle. More precisely, their frames act as one frame. Field lines that are not, or not completely,

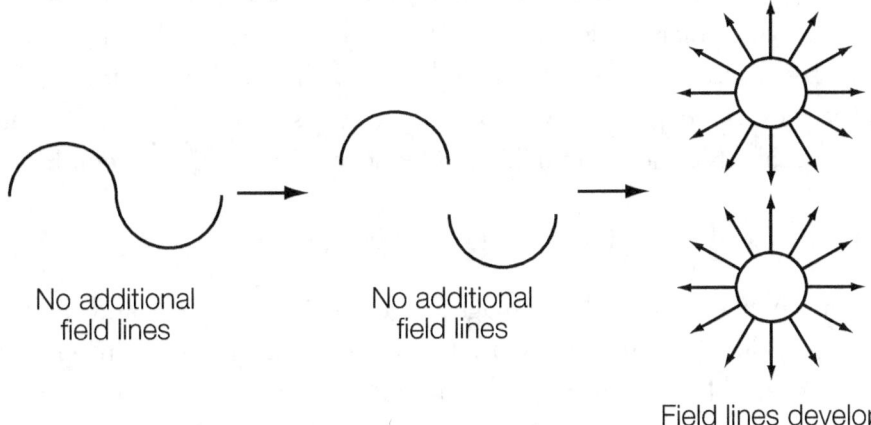

No additional
field lines

No additional
field lines

Field lines develop

Figure 3.40 *The field lines emanating from electrically charged leptons, quarks, as well as neutrinos do not form until such particles are fully formed.*

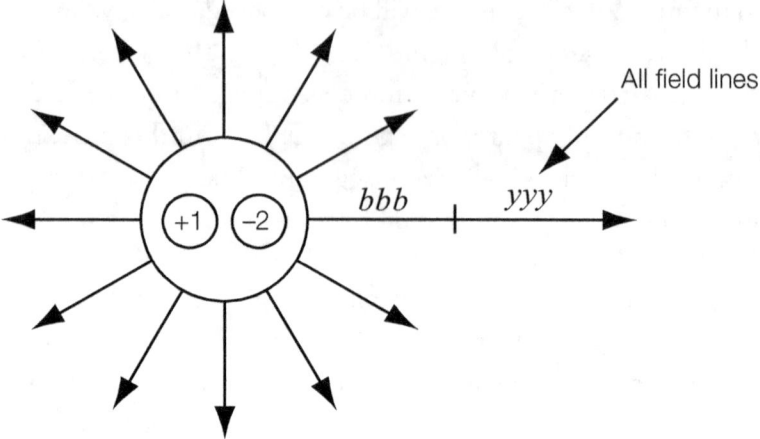

Figure 3.41 *When particles bond with each other, their frames act as one frame.*

canceled out during the bonding will dictate the temporal/electrical nature of the common frame. For example, if a particle with a charge of +1 bonded with a particle with a charge of −2, the second particle would have field lines not involved in the bonding, and the common frame would behave like one negatively charged particle (*Figure 3.41*).

When charged particles combine to form a neutral composite particle, the surrounding field lines emanating from the individual particles disappear. Consider two generic electrically charged particles, one with +1 charge and the other with −1. As they bond with each other, the field lines linking them shrink, as described previously. This action sends feedback to each frame, inhibiting its ability to generate field lines along other parts of its circumference.

Maintenance of the Electromagnetic Force

The maintenance of the electromagnetic force in both QM and TET has everything to do with photons—virtual and real. As long as photons exist, the force exists. In TET, of course, this ultimately relates to the ingredients of photons: electric field lines and temporal particles, particularly blue lines and t^+ particles, because as noted earlier, the t^+/t^- process of blue lines supplies all (or most) of the t^- particles for maintaining the yellow components of electric field lines and also helps bring about a full magnetic force between two objects.

Although QM has processes built into its framework that are equivalent to TET's t^-/t^+ process for the creation of virtual and real photons, it does not have a process equivalent to TET's t^+/t^- process. QM regards the mere existence of the photons as "good enough" for the maintenance of the force, thus focusing only on its energy-borrowing process (t^-/t^+) to create the photons. In TET, the t^+/t^- process not only maintains electric field lines and allows magnetism in its fullest form to occur, it also eliminates excess energy. Because QM has no process like this, it produces infinite energy results. From the perspective of TET, this energy is t^+ energy. At first glance, it might seem that the infinities would be eliminated by the QM concept of energy borrowed from the vacuum being quickly returned. However, although energy is borrowed and paid back quickly in QM, there seems to be, from moment to moment, a constant presence of all of the vacuum's energy within the field of photons, which mathematically looks like infinity, increasing their mass. This is because, when borrowing energy from the vacuum, there is no way for QM to say to it, "I only want this much and no more," and the borrowing process is a continuous one. As QM does not link the payback of the energy to the fundamental progression of time, the infinite energy seems to always be present.

The reason TET does not have infinite energy is because as energy is added to a system over a fundamental, indivisible moment of time, it is also removed over that same moment, such that, from moment to moment, the energy level of the system never changes. The energy being taken away is not necessarily the energy being borrowed during that moment, but the amount of energy being borrowed and removed is the same. *Figure 3.42* is a schematic of the process for a virtual photon and an electron. Each arrow in the figure represents the passing of a fundamental moment of time.

Exterior space is represented by the t^0 on the left. Arrow 1 represents the emission of t^+ energy from exterior space into interior space. The t^+ particle between arrows 1 and 2 has one foot in exterior space and the other in interior space. Arrow 2 represents the t^+ particle becoming a full-fledged part

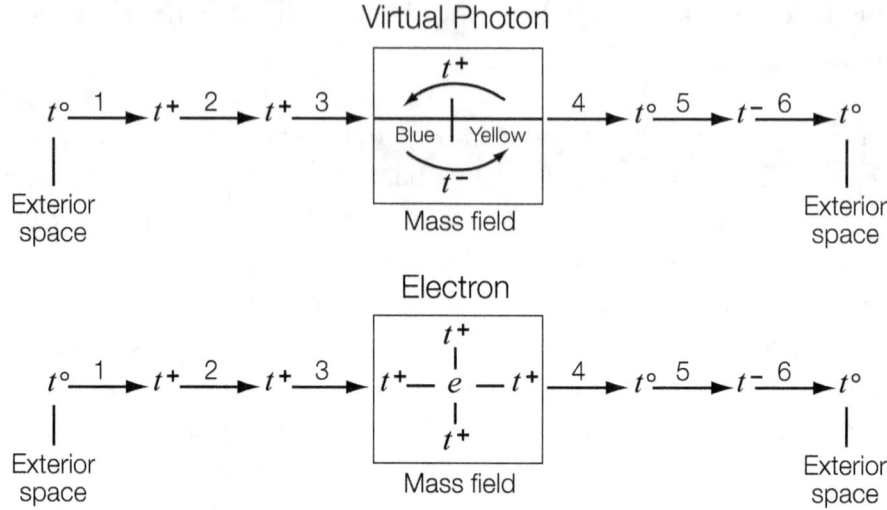

Figure 3.42 *In TET, as energy is added to a system over a fundamental, indivisible moment of time (arrow 3), it is also removed over that same moment (arrow 4), such that the level of energy does not appear to change with simple time progression.*

of the field of t^+ particles housed in interior space. Arrow 3 represents the t^+ particle from interior space transitioning into the photon and electron's mass fields, denoted by the square boxes. The t^+ particle between arrows 2 and 3 has one foot in interior space and another in the mass fields of the photon and electron. Arrow 4 represents a t^+ particle transitioning out of the mass fields into interior space. The t^0 particle between arrows 4 and 5 has one foot in the mass fields and one in interior space. To understand this transition, recall that blue field lines attract and form bonds with t^+ particles and repel t^- particles. When a t^+ particle transforms into a t^0 particle, it becomes more negative. If the yellow component of a photon already has the t^- it needs, any extra t^- will be ejected by the system. Arrow 5 represents the temporal particle (t^0) becoming once again a full-fledged part of the temporal particles housed in interior space, but this time as a t^- particle. Arrow 6 represents the absorption of the t^- by exterior space.

The conversion of t^0 to t^+, t^+ to t^0, t^0 to t^-, and t^- to t^0, each represents the smallest unit of time. It is only when one of these transformations has occurred that one fundamental moment of time can be said to have passed. Arrows 2 and 3 also represent fundamental moments of time, because they represent movement of a t^+ particle into a mass field, which in TET can happen only as another t^+ particle begins to vacate the mass field, which occurs at arrow 4 (t^+ to t^0). As arrow 3 begins to add energy to the systems over a single moment of time, arrow 4 begins to take energy away over that same moment. Stated another way, as energy begins to be borrowed at arrow 3 over a fundamental moment of time, it begins to be paid back at arrow 4 during that same time, as the processes are occurring simultaneously, such that the mass of the systems remains constant as time progresses.

Arrow 3 ultimately reduces to the t^0 to t^+ process, which is of course part of the t^-/t^+ process of exterior space. Arrow 4, t^+ to t^0, is part of the t^+/t^- process of the photon and electron. But as with the subprocesses (t^0 to t^+ and t^+ to t^0), the full processes of t^-/t^+ and t^+/t^- occur simultaneously, such that the energy of the systems does not appear to change over time, even though energy is constantly being borrowed and paid back. Only the temporal particles that represent the mass of the photon and electron appear to linger, which is not an infinite value. Note that once the photon and electron return temporal energy to interior space, which they do as t^-, they have satisfied their obligation, because it was from interior space that they took energy, as t^+. It is up to exterior space to take back the energy it gave to interior space. Exterior space gives interior space energy as t^+ and takes energy away as t^-.

From the perspective of TET, QM obtains infinite energy values because, although it links the borrowing of energy to the fundamental progression of time, it does not link the payback of that energy to this time. In QM, the vacuum loans energy for the creation of a system or "cloud" of virtual photons around a stationary charge, for example, as time progresses, but it is up to the photons to return the energy as quickly. However, in QM, photons do not do this, nor do other particles in the theory. For this to happen, QM would have to incorporate a time process that runs opposite to that of its energy-borrowing process—a time process that returns energy—the t^+/t^- process from the perspective of TET. Thus, from moment to moment in QM, borrowed energy seems to always be present. In that theory, the energy must be paid back over some time interval, albeit short if a large amount of energy is borrowed, longer if a small amount of energy is borrowed, but before that interval is up, the energy

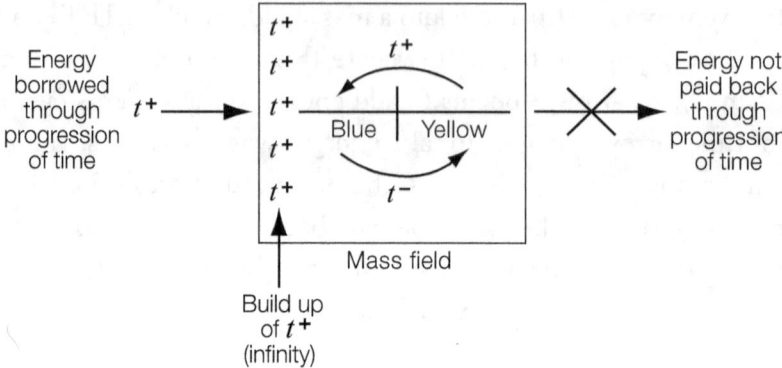

Figure 3.43 *Shown is the build up of infinite energy in QM from the TET perspective.*

can essentially just "hang out" (*Figure 3.43*). As the borrowing process is a continuous one and the payback is not linked to the fundamental progression of time, infinite energy seems to be ever present. In TET, energy is borrowed and paid back to the vacuum literally as time progresses on a fundamental basis. Energy never appears to accumulate simply with the progression of time. Energy is paid back more quickly in TET than in QM.

In order to eliminate the infinite energy values in QM, the ad hoc process of renormalization must be invoked. Earlier, this process was used to hide the negative energy vacuum by laying a positive layer of energy on top of it. In the current instance, it is used to place a layer of infinite negative energy on top of the infinite positive energy of the photons in QM. From the perspective of TET, the laying of positive energy over the negative energy vacuum was seen as a benign procedure because in TET, there is a positive energy vacuum that is equivalent to that layer of positive energy. Similarly, from the perspective of TET, the laying of an infinite layer of negative energy over the infinite positive energy of QM's photons is also a benign procedure. It is equivalent to recognizing the t$^+$/t$^-$ process, in that the infinite positive layer of energy represents t$^+$ and the infinite negative layer of energy represents t$^-$. In this sense, the t$^-$ is viewed as mathematically canceling the t$^+$. In TET, of course, t$^+$ is transformed into t$^-$ and t$^-$ is absorbed by exterior space. Through either mathematical cancellation or the processes of transformation and absorption, however, excess energy is removed—and more quickly than through the normal energy payback mechanism of QM—eliminating the infinite results (*Figure 3.44*).

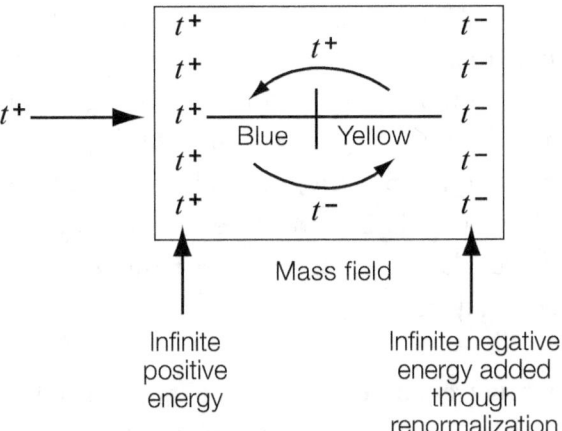

Figure 3.44 *Shown is the application of renormalization from the TET perspective. Infinite negative energy is added through renormalization to mathematically balance the energy level, leaving only normal mass in the mass field.*

The Weak Force and Temporal Energy Theory

The weak force is an assortment of phenomena that can affect all QM particles, but which is ultimately produced by those that have slightly polarized segments in their frames, mainly neutrinos and quarks. This section will focus on the weak force as brought about by neutrinos. In addition to their slightly polarized segments, quarks also have fully polarized segments. The weak force as associated with quarks will be discussed in more detail in the section on the strong force and in chapter 8.

One aspect of the weak force as brought about by neutrinos is similar to the electric force, except that the field lines involved—weak field lines—are temporally polarized only slightly. As temporal polarization and electric charge are the same thing, weak field lines technically have a slight electric charge. In general, the weak force can be thought of as a weak form of the electric force. However, neutrinos are considered to be electrically neutral. The reason for this in TET has to do with the field lines emanating from the particles. Unlike electric field lines, whose only limit in terms of how far they reach is the extent of interior space itself, the field lines emanating from neutrinos are short ranged. Recall that an electron's electric field (really the electric field of any electrically charged particle) is a property of interior space. The electron's presence simply jumpstarts a process that really occurs in, by, and through interior space itself, the vacuum. The neutrino, not being fully polarized, does not have enough temporal "juice" to get this process fully going. Interior space is slightly polarized close to the particle, in similar fashion to the polarization of the particle itself, but the polarization process diminishes not far from the particle. It does not have the same reach that, for example, an electron does, and thus appears electrically neutral.

The field lines emanating from an electrically charged particle can reach over to the field lines emanating from a neutrino to cause an interaction. However, the neutrino field lines are not fully polarized, whereas those of the electrically charged particle are. As a result, there would never be a full attraction or repulsion between an electrically charged particle and a neutrino, just a wiggling or jiggling of the particles. Neutrinos, however, can fully attract or repel one another, in similar fashion to electrically charged particles, because their field lines are on equal footing, but they have to be in close range (*Figure 3.45*). Considering that there is some charge on a neutrino but it is not exactly like electric charge, you could say that neutrinos have weak electric charge or simply weak charge. Thus, an electron neutrino would have a weak charge of −1, whereas an electron would have an electric charge of −1. Weak charge can be viewed as a very diluted form of electric charge in TET.

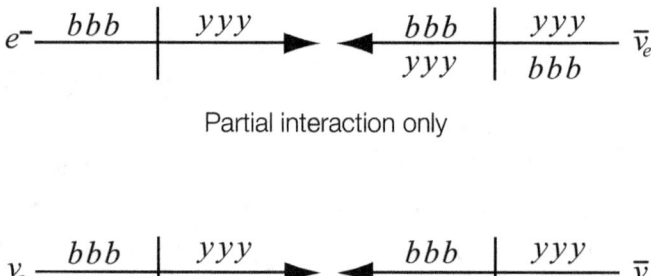

Partial interaction only

Full interaction (attraction) possible

Figure 3.45 *An electrically charged particle and a neutrino are able to interact only partially because their field lines are not on equal footing. Neutrinos are able to interact with each other fully at close range.*

Note that the field lines emanating from a neutrino and the field lines emanating from an electron are all composed of the same stuff, or non-stuff—interior space. It is just that the neutrino's field lines are not fully polarized, whereas the electron's field lines are. Note also that as the frame of an electron is just a blue electric field line wrapped back on itself, the frame of an electron neutrino is just a bluish-green weak field line wrapped back on itself. Likewise, as the frame of a positron is a yellow electric field line wrapped back on itself, the frame of an anti–electron neutrino is just a yellowish-green weak field line wrapped back on itself.

Thus far, the weak force phenomena can be summed up as (1) the full attraction or repulsion of neutrinos (including anti-neutrinos) through weak field lines at close range and (2) the wiggling and jiggling of a neutrino and an electrically charged particle, as the two come near each other and the weak field lines of the neutrino interact with the electric field lines of the electrically charged particle. Another weak force phenomenon is the transformation of particles from one type to another, which may occur through changes in the polarization of the segments within the frames of particles, to bring about stability to a particle. The reason this is a weak force phenomenon is because particles composed of weak (including partly weak in the case of quarks) frames are the most energetically favorable for carrying out the transformation. In order for the segments within the frame of a particle to shift their polarization, the particle must partner with another particle whose segments will be shifted oppositely. Particles with at least one weak segment within their frames require less energy to shift their polarization than a particle with a solely electric frame. Their weak nature makes them more amenable to a polarization change than a fully blue or yellow particle. The ease by which they can be changed can be used by another particle to change its own polarization. Therefore, particles often partner with neutrinos to shift their polarization, and of course, in so doing, the neutrinos also shift their polarization. The result is two new types of particles.

The polarization shift, also called repolarization, follows the same blue-yellow format: That is, one particle in the partnership is blue shifted, while the other is yellow shifted. Also, neutrinos, which contain bluish-green segments, can

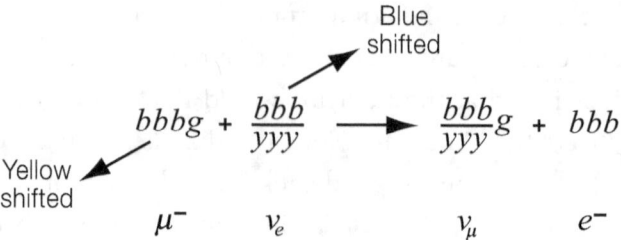

$$bbbg + \frac{bbb}{yyy} \longrightarrow \frac{bbb}{yyy}g + bbb$$

Blue shifted

Yellow shifted

μ^- v_e v_μ e^-

Figure 3.46 *When a muon interacts with an electron neutrino, the muon can yellow shift, changing into a muon neutrino, and the electron neutrino can blue shift, changing into an electron.*

only be blue shifted, whereas anti-neutrinos, which contain yellowish-green segments, can only be yellow shifted. The nature of the unstable particle being transformed (to gain stability) will dictate whether it will interact with a neutrino or an anti-neutrino. For example, a muon is a very unstable particle and thus wishes to transform into something more stable. In order to do this, it partners with an electron neutrino. Through this partnership, the muon is able to yellow shift and become a muon neutrino and the electron neutrino blue shifts and becomes an electron (*Figure 3.46*). The reason the muon does not partner with an anti–electron neutrino is because such a particle can only yellow shift itself, but the muon cannot blue shift because it is already fully blue. Thus, no reaction occurs, and the muon remains in its unstable state.

The neutrino or anti-neutrino that a particle partners with in these transformation processes can be of any type depending on the energy of the system: the electron type, muon type, or tauon type. They may preexist in space or they may be manufactured by the system if enough energy is available. Recall that when the field line system of an electrically charged particle is static, it is simply referred to as an electric or electrostatic field. When the field line system moves, it develops a magnetic nature and is referred to as an electro-magnetic field, having the form of waves. This also occurs with weak field lines emanating from particles. In TET, a static system of weak field lines is called a weak field. When the lines move, it is called a weak-magnetic field. In TET, the weak-magnetic force is of course simply a weak form of the electromagnetic force. (The link between the weak and electromagnetic forces has been known for some time, and together, they have been referred to as the electroweak force.) As an electron/positron pair can arise from an electromagnetic field, an electron neutrino/anti–electron neutrino pair can arise from a weak-magnetic field, with one of the neutrinos perhaps going on to cause a particle transformation (*Figure 3.47*).

Consider the transformation of a muon into a muon neutrino: The instability of the muon will cause it to jostle around wildly, which in turn will cause its electromagnetic field to fluctuate greatly. Under such high stress, an electromagnetic field will fluctuate into a weak-magnetic field. To help in understanding this, consider two weak individuals, with neither having the strength to lift a 10-kilogram box off a floor. Under further examination, you realize

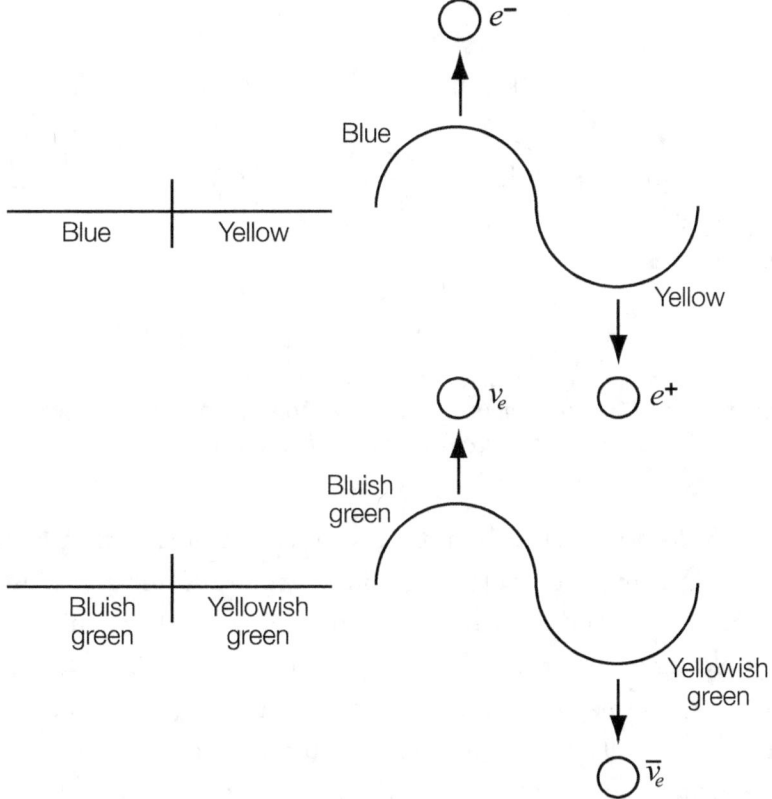

Figure 3.47 *A weak field line is analogous to an electric field line; it is a weak version of an electric field line. As an electron/positron pair can arise from an electromagnetic field, an electron neutrino/anti–electron neutrino pair can arise from a weak-magnetic field.*

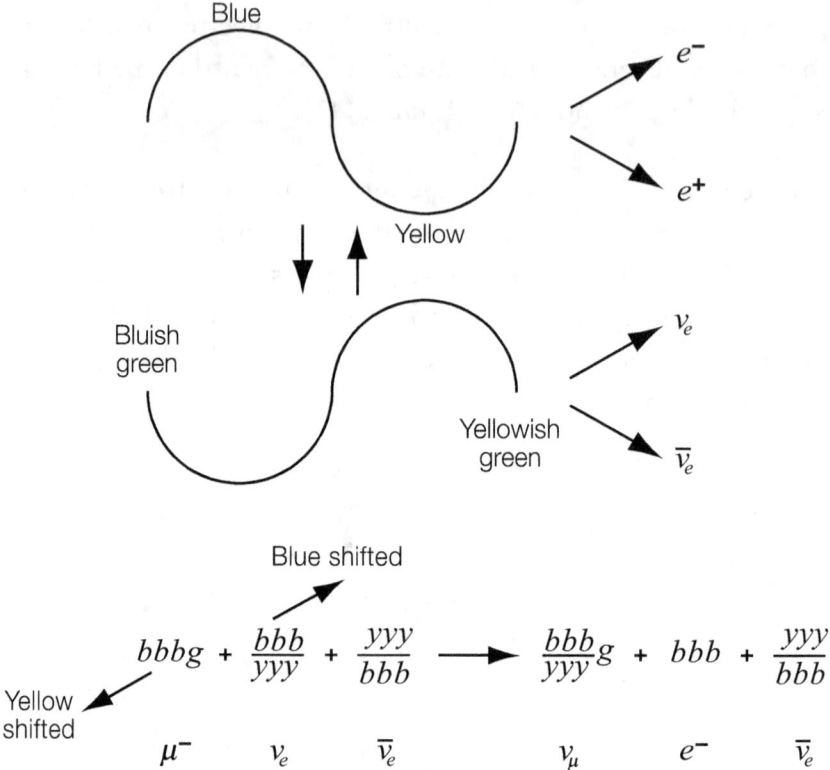

Figure 3.48 *An electromagnetic field can transform into a weak-magnetic field. An electron neutrino/anti–electron neutrino pair can then arise from the weak-magnetic field, with the electron neutrino interacting, in this case, with a muon.*

that person A does not have the muscles to lift the box. Person B, however, does, but has just finished a very strenuous exercise routine and is exhausted. Thus, neither is able to lift the box. Person A is like a weak-magnetic field, and person B is like an electromagnetic field that has transformed into a weak-magnetic field because of high stress. As noted earlier, as an electron and positron can be created from an electromagnetic field, an electron neutrino and an anti–electron neutrino can be created from a weak-magnetic field, with the electron neutrino interacting with the muon as described above (*Figure 3.48*).

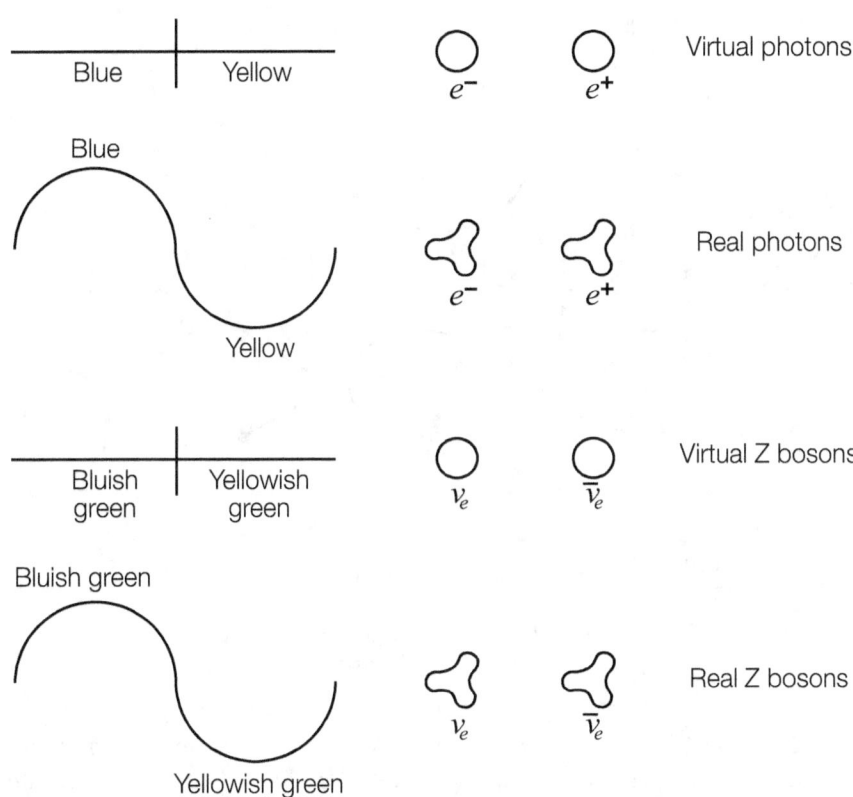

Figure 3.49 *Virtual and real Z bosons are analogous to virtual and real photons.*

Another way of describing the creation of an electron and positron is that they arise from a neutral photon. The electron neutrino and anti–electron neutrino in like manner arise from a neutral particle referred to as the Z boson. That is, whereas a neutral photon is composed of blue and yellow lines and temporal particles, a Z boson is composed of bluish-green and yellowish-green lines and temporal particles. As a neutral photon is an electrically negative and positive photon pair, a Z boson is a weakly negative and positive Z boson pair. Also, as the electron and positron are electrically charged photons separated from each other, the electron neutrino and anti–electron neutrino are weakly charged Z bosons separated from each other. The blue "side" of all neutrinos qualifies them as full magnets, like electrons, as the blueness results in a loss of t^+ energy, a necessary phenomenon for full magnetism in TET. Thus, as

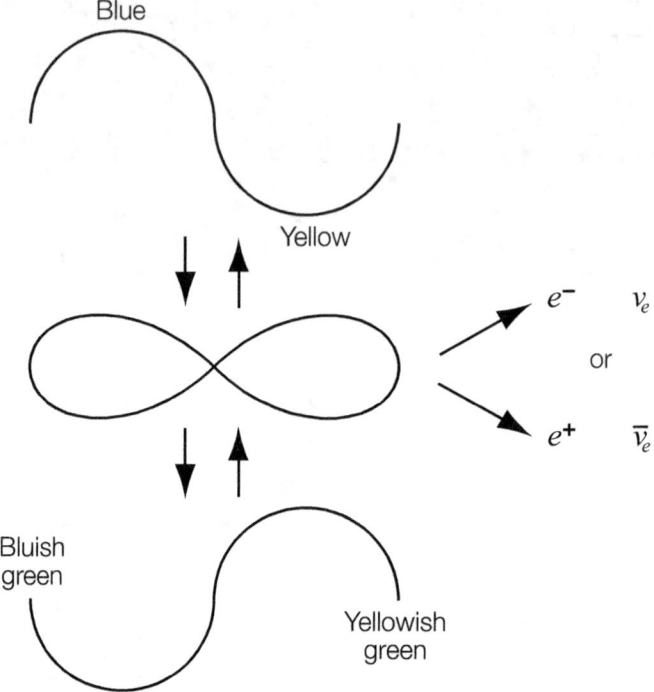

Figure 3.50 *In TET, when an electromagnetic field is transformed into a weak-magnetic field, a 2-loop standing wave may develop, similar to what might develop when a string that is bound on both ends is made to vibrate. This standing wave may appear as a large particle with ultra-high mass. It is likely the "Z boson" found in laboratory experiments.*

photons are involved in electric and magnetic forces, or the electromagnetic force, Z bosons are involved in weak and magnetic forces, or the weak-magnetic force. When there are no disturbances in a weak field line or the frame of a neutrino, a virtual Z boson can be said to exist. When there are such disturbances, a real Z boson can be said to exist (*Figure 3.49*).

When a Z boson is created in the laboratory, it appears to have an ultra high mass. From the perspective of TET, this occurs because in the laboratory the Z boson is being created through an electromagnetic field. As indicated above, for this to happen the electromagnetic field has to be put under high stress. As a result, a 2-loop standing wave likely develops, similar to what might develop when a string that is bound on both ends is made to vibrate (*Figure 3.50*). From the standpoint of TET, it is the energy signature of this standing wave—this pseudo-structure fluctuating between the photon (electric) and the Z boson (weak) states—that is found when an attempt is made to create a Z boson in the laboratory. With so much activity, the standing wave attracts many t^+ particles and, in and of itself, looks like one very large, massive particle. It may give rise to a very high energy electron/positron pair or a very high energy electron neutrino/anti–electron neutrino pair.

Sometimes the 2-loop standing wave may split into two single-loop standing waves, which would still be fluctuating between the electric and weak states, before degrading and perhaps causing a particle transformation, as described above. From the TET perspective, the energy signatures of these two separate loops are what you would find if you searched in the laboratory for what are called the W^- and W^+ bosons (*Figure 3.51*). In TET, however, an actual W boson is simply a neutrino and sometimes a quark. It is the particle against which an unstable particle is leveraged to change the polarization of the

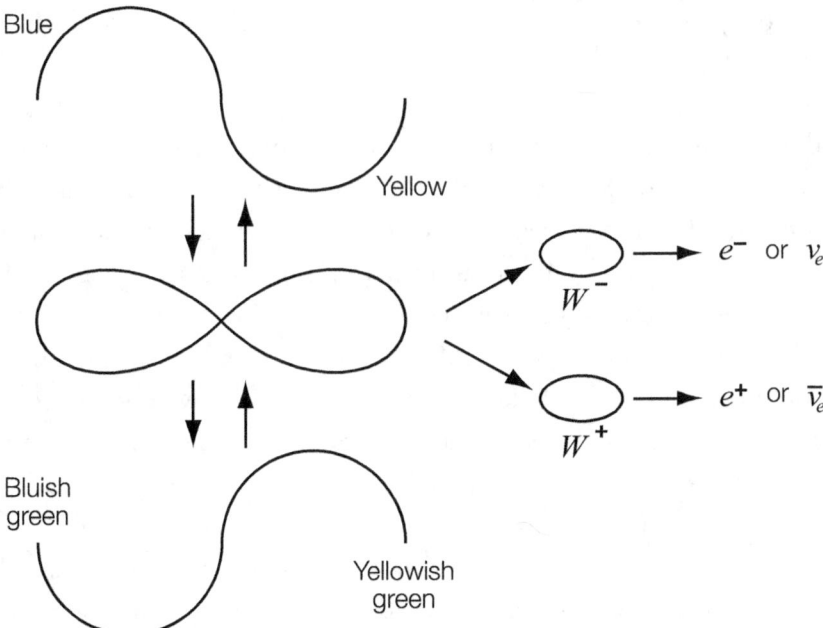

Figure 3.51 *The 2-loop standing wave may split into two single-loop standing waves, which would still be fluctuating between the electric and weak states. From the TET perspective, the energy signatures of these loops are the W^- and W^+ bosons found in laboratory experiments.*

unstable particle (and itself as a result) to bring about stability in that once unstable particle. In TET, the W boson is not the single-loop standing wave structure, just as the 2-loop standing wave structure is not the Z boson, but it is what likely would be found if you looked for one in the laboratory.

Z bosons are thought to be the particles responsible for the momentum changes when, for example, an electrically charged particle passes by a neutrino and the two particles appear to jostle one another. This is also the case in TET, except that as with photons regarding the electromagnetic force, only the ingredients of Z bosons, weak field lines and temporal particles, are needed to bring about this reaction. Neither a real Z boson nor the standing wave "Z boson" is needed. If they exist, that is fine; if they do not, that is fine too. The W bosons are thought to be responsible for particle transformations, which is the case in both QM and TET. However, again, in TET, W bosons are neutrinos or quarks. Although momentum changes can be caused by virtual or real Z bosons, particle transformations require real W bosons, that is, moving neutrinos or quarks in TET. Stationary neutrinos or quarks would likely not lead to particle transformations.

Note that there are many instances in which neutrinos are created under energy conditions too low to create standing-wave Z or W bosons. Earlier it was stated that it takes a great deal of energy to change an electromagnetic field into a weak-magnetic field (and thus into neutrinos). However, this is a "pure" electromagnetic field—that generated, for instance, from an electron or positron. The electromagnetic field generated by a muon, anti-muon, tauon, or anti-tauon is a little less pure because they have one or more green segments in their frames, whereas an electron and positron do not. Although on average the electromagnetic field generated by, for example, an electron and muon are the same, the presence of the green segment within the frame of the muon lowers the energy threshold needed for changing its electro-magnetic field into a weak-magnetic field. This is because although the green segment is essentially true green, it fluctuates slightly between bluish green and yellowish green, with the average between the two states being true green. Muons, anti-muons, tauons, and anti-tauons have some weakness about them, albeit transient, due to their green segments, and thus, their electromagnetic field lines are able to transform into weak-magnetic field lines and produce neutrinos by way of a lower amount of energy than it would otherwise take, circumventing the creation of standing wave W or Z bosons. Particles with true weak natures, like neutrinos and quarks, whose field lines are naturally weak—although only partly so in the case of quarks—have an even lower energy threshold for producing neutrinos.

Sterile neutrinos produce no weak field lines or, at best, weak field lines with only transient existence. This is because their frames are true green, having no permanent polarization. Sterile neutrinos are massless, and their spins involve random spatial motion. (Thus, they do not behave like little magnets when stationary the way, for example, electrons and electron-neutrinos do.) Because of this, interactions with sterile neutrinos are not likely. With their lack of radiating field lines, true-green frames, lack of mass, and random spins, they blend in very well with the vacuum, making it difficult to distinguish them from what would be referred to as empty space. They should possess momentum, however, and be able to alter the position and momentum of another particle they bump into.

Maintenance of Weak Field Lines

The ultimate process maintaining weak field lines is the t^+/t^- process of the bluish-green field lines, as the t^+/t^- process of blue lines maintains electric field lines (*Figure* 3.52). Although the bluish-green lines process time in both directions, there is a slight favoring of the t^+/t^- process. Some of the discarded t^- from the bluish-green lines is used by the yellowish-green lines to help them maintain their polarization. Although the yellowish-green lines, requiring little t^- compared to fully yellow lines, can likely extract an adequate amount of t^- from interior space, it is energetically more favorable to simply use the t^- their bluish-green partners discard. Neutrinos also maintain their charge through their t^+/t^- process. Anti-neutrinos use the t^-/t^+ process; being separated from their sister particles but needing relatively little t^- for their survival, anti-neutrinos successfully extract some of the t^- they need from interior space.

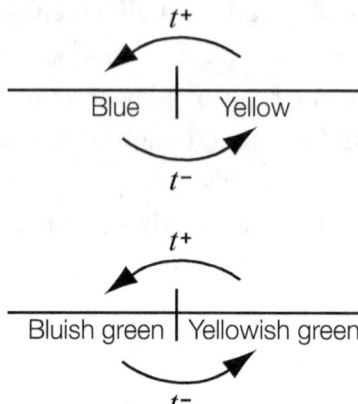

Figure 3.52 *The ultimate process maintaining weak field lines is the t^+/t^- process of the bluish-green field lines, as the t^+/t^- process of blue lines maintains electric field lines.*

In a certain way, it might seem that weak field lines do not need to interact with exterior space. That is, the bluish-green lines require little t^+, and the yellowish-green lines require little t^-. It might seem that they could just extract what they need from interior space, transfer the energy between themselves as needed, and return it to interior space to be used again. The problem is that any system involving t^- will constantly lose energy because of exterior space's strong affinity for it. Therefore, it will require an influx of energy as well, which is supplied by the steady stream of t^+ from exterior space (*Figure* 3.53).

Maintenance of Weak-Magnetic Force

The maintenance of the weak-magnetic force occurs in a similar fashion to the maintenance of the electromagnetic force. From the standpoint of TET, this means through 1) the t^-/t^+ process regarding the creation of virtual and real Z and W bosons and 2) the t^+/t^- process regarding the maintenance of weak field lines and frames. Again, however, although QM has processes equivalent to t^-/t^+ built into it to create virtual and real Z and W bosons as it does virtual and real photons, it has no process equivalent to t^+/t^-. QM regards the mere existence of the bosons as "good enough" for the maintenance of the force, thus focusing only on its energy-borrowing process (t^-/t^+), as this process

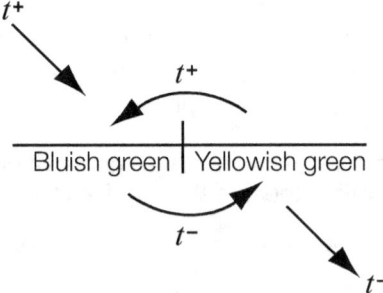

Figure 3.53 *Weak field lines continuously borrow energy in the form of t^+ from exterior space and repay that energy in the form of t^-.*

supplies the energy to create the bosons. As noted for the electromagnetic force, focusing only on the t^-/t^+ process leads to infinite energy results, as t^+ is borrowed as time progresses (through the t^-/t^+ time process) but is not paid back as quickly (through the t^+/t^- time process). Thus, the ad hoc procedure of renormalization must be used to eliminate the infinite energy, which is equivalent to recognizing the t^+/t^- process, as described earlier.

The Strong Force and Temporal Energy Theory

In TET, the strong force is considered to be a combination of the electric and weak forces. The combination makes this force stronger than either of the other two. This force is associated with quarks and, like the electric and weak forces, is brought about through field lines, which will be called strong field lines, and their associated temporal particles. The field lines have actually already been introduced in the section "Important Relationship Between Particle Frames and Field Lines." They are the mixed field lines emanating from quarks, having both electric and weak qualities, like the quarks themselves. Only quarks experience the strong force. Quarks combine with other quarks through this force to form composite particles called baryons and mesons. Baryons are combinations of three quarks or three anti-quarks; mesons are quark/anti-quark pairs of different types. An example of a baryon is the proton; it is composed of two up quarks and one down quark. An example of a meson is a positive pion, which is composed of an up quark and an anti-down quark.

Table 3.9 *Rules for Applying Overlap Method to Field Lines*

I. For full interaction, field lines must be of same species (electric with electric, weak with weak, strong with strong).

II. For strong field lines, in which segments of different species may interact within the overlapping field lines,

    A. Identically polarized segments cancel first.

    B. Similarly polarized segments cancel second (for example, blue with bluish green).

    C. Oppositely polarized segments interact last (electric-electric interactions, followed by electric-weak, followed by weak-weak).

With regard to the electric and weak forces, opposite charges attract, and like charges repel. The same principles apply to the strong force. Quarks have both electric and weak charge, and depending on the types of quarks involved in an interaction and the dynamics between them, the electric force may dominate or the weak force, or a combination of the two. Interestingly, once quarks enter into stable combinations, pulling them apart is nearly impossible. Actually, the amount of energy needed to pull them apart is enough to create new quarks, such that you always end up with one stuck to another. No individual free quarks have ever been discovered.

To understand why quarks combine, however, it is useful to consider them as free particles, which they were in the high-energy environment of the early universe. Although there are three generations of quarks—see *Table 3.8*—those in the second and third generations generate the same types of strong field lines as those in the first generation, so for brevity, this section will focus only on those in the first generation: the up quark, down quark, and their anti-particles. Also, it is helpful to use the overlap method described in the section on the electric force. However, there are rules of the method that must be considered (*Table 3.9*). The first rule applies to field lines in general, and says that for there to be a full interaction (attraction or repulsion), the field lines interacting must be of the same "species." As discussed earlier, for example,

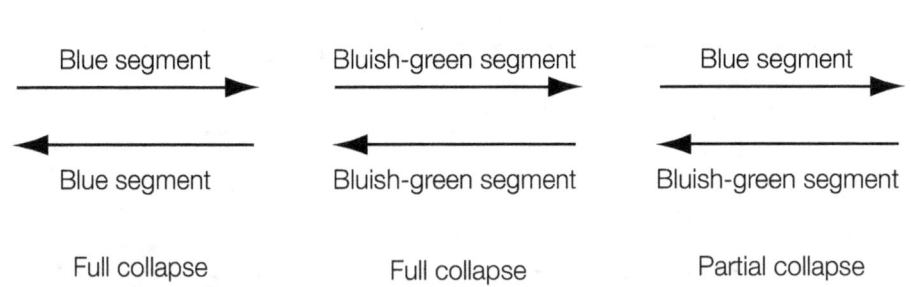

Figure 3.54 *If the overlapping field line segments are of different species (for example, blue and bluish green), only a partial collapse may occur.*

an electric field line meeting a weak field line only results in a wiggling and jiggling of the particles generating those field lines. The second rule, which applies only to strong field lines, is that identically polarized segments (blue and blue, yellow and yellow, bluish-green and bluish-green, or yellowish-green and yellowish-green) within overlapping lines must be considered to cancel each other first, followed by similarly polarized segments (blue and bluish green or yellow and yellowish green). After this, oppositely polarized segments should be considered to interact.

Note that unlike in electric attraction, there is not necessarily a full zeroing out of the field line segments in strong force attraction. For example, consider a blue segment being generated to the right going against a bluish-green segment being generated to the left. If both segments were fully blue (or for that matter if both were bluish-green), a full collapse would ensue from the overlap; instead, only a partial collapse occurs. Whether two quarks will attract or repel depends on the net effect of all the different attractions and repulsions that may be happening between them (*Figure* 3.54 and 3.55).

For example, despite having the same electric and weak charge, two down quarks are likely to bond with each other. There is a slight repulsive force between them, but there is a slightly greater attractive force (Figure 3.56). An up quark and a down quark are even more likely to bond than the two down quarks, owing to a strong attraction within the overlapping field lines (*Figure* 3.57). Two up quarks are likely to repel one another (*Figure* 3.58). However, if an additional up quark joins in the bond between an up quark and a down quark, a stable composite particle is created, which of course is the proton.

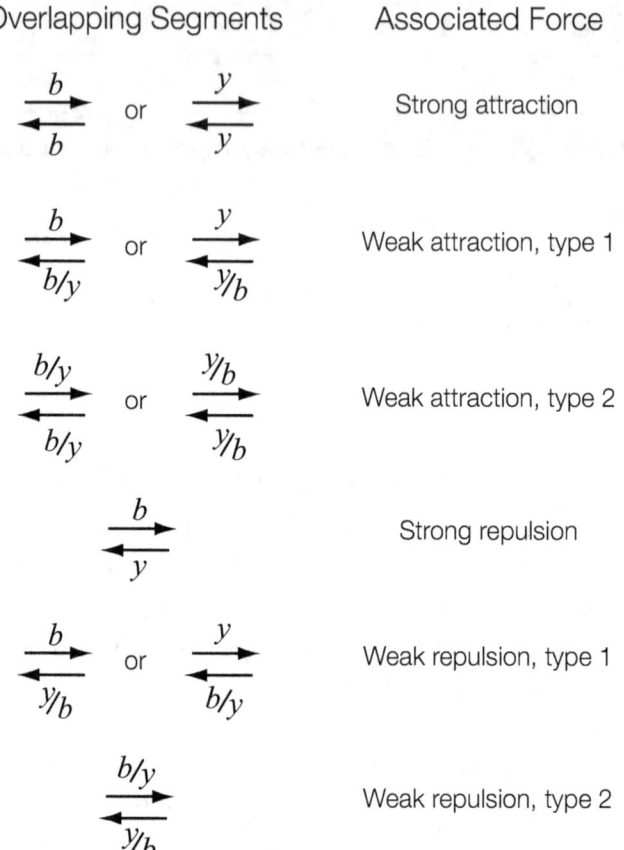

Figure 3.55 *Attractions and repulsions of differing strengths can occur depending on the nature of the overlapping segments. In the figure, which segment is pointing to the left and which is pointing to the right is arbitrarily assigned. Type 1 weak attraction or weak repulsion is greater than type 2 weak attraction or weak repulsion.*

The principal reason an up quark combines with an up quark/down quark combination to form a proton is because of the weak force component of the strong force. Each up quark has –1/3 weak charge (in addition to its electric charge of +2/3). The down quark has a +2/3 weak charge (in addition to its –1/3 electric charge). With there being two up quarks in a proton, there are two units of –1/3 weak charge, and together, these balance the +2/3 weak charge of the down quark (*Figure 3.59*).

Even when combined in a proton, the two up quarks repel each other, which under most circumstances would tear such a composite particle apart.

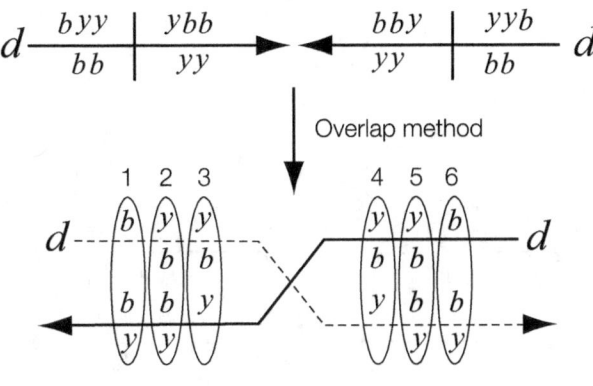

1, 3, 4, 6 = Weak attraction, type 1

2, 5 = Weak repulsion, type 2

Figure 3.56 *Two down quarks are likely to bond, as there are four type 1 weak attractions compared with two type 2 weak repulsions.*

However, when the proton is pulled apart, an internal attractive electric force builds up within the particle. The further the quarks composing the proton are separated, the greater the magnitude of this electric force, such that it becomes increasingly more difficult to pull the quarks apart (*Figure 3.60*), hence the term strong force. As noted above, the amount of energy needed to

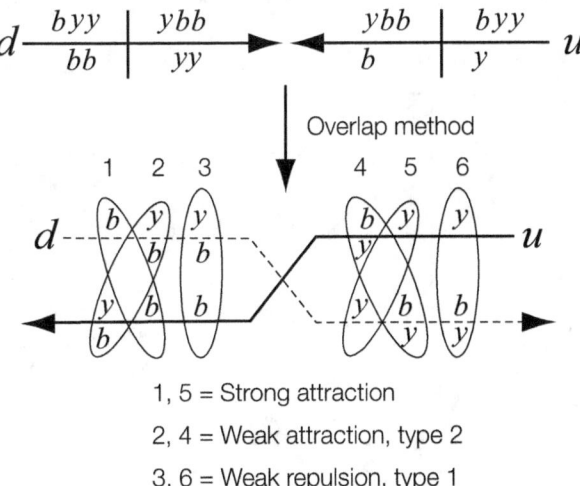

1, 5 = Strong attraction

2, 4 = Weak attraction, type 2

3, 6 = Weak repulsion, type 1

Figure 3.57 *Down and up quarks are likely to bond. There is a strong attraction. And although they are different grades, the weak attraction and weak repulsion just about cancel each other.*

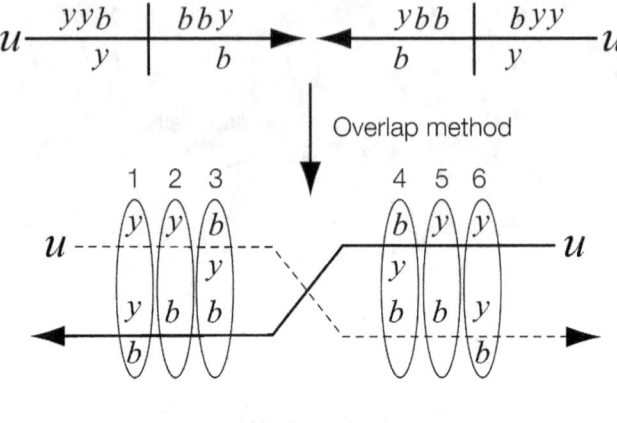

1, 3, 4, 6 = Weak attraction, type 1

2, 5 = Stong repulsion

Figure 3.58 *Two up quarks would likely repel each other. Even with the weak attractions, the strong repulsions would likely drive the quarks apart. At best, the interactions might balance each other.*

pull the quarks apart is enough to create new quarks, such that there are always quarks attached to other quarks (*Figure 3.61*). Thus, because the attraction between quarks increases the further they are separated, the repulsive force between the two positively charged quarks within a proton is not enough to break up the particle. Once the weak force component of the

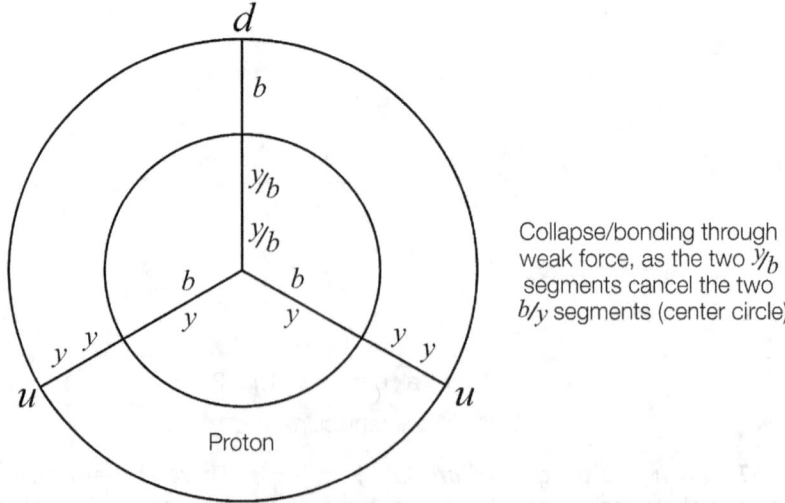

Collapse/bonding through weak force, as the two y/b segments cancel the two b/y segments (center circle)

Figure 3.59 *The proton (uud) is held together primarily through the weak force.*

strong force establishes a connection between the quarks, that connection becomes essentially impossible to break.

As an up quark can join an up quark/down quark combination to form a proton, an up quark can also join a down quark/down quark combination to form a neutron. In contrast to the proton, the principal reason an up quark combines to form a neutron is because of the electric force component of the strong force. Like all up quarks, the up quark in a neutron has an electric charge of +2/3, and the two down quarks each have an electric charge of −1/3. Thus, the neutron is held together mainly because of electric forces (*Figure 3.62*).

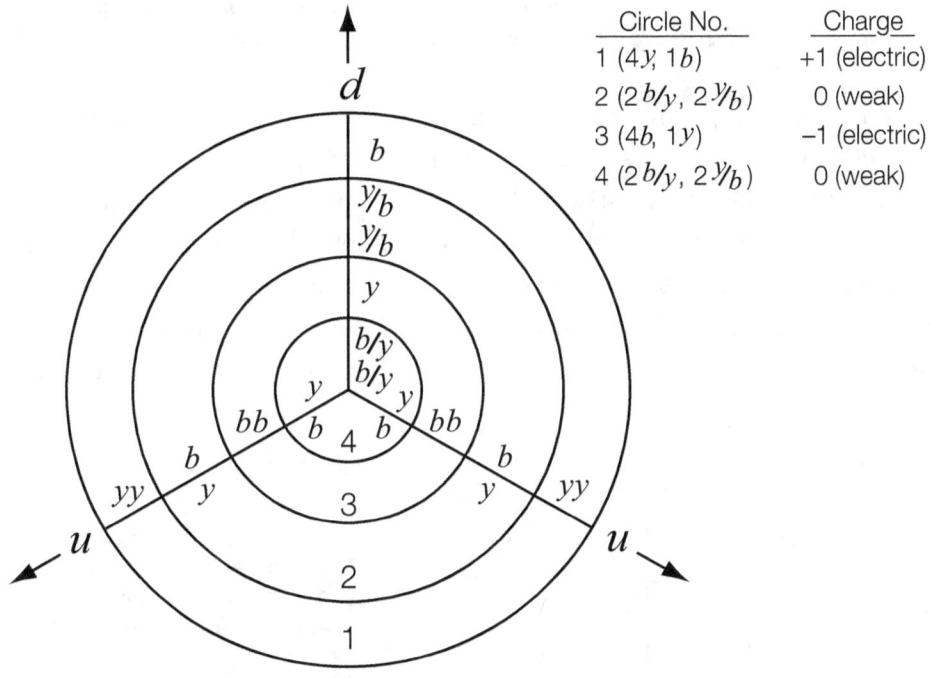

Circle No.	Charge
1 (4y, 1b)	+1 (electric)
2 (2b/y, 2y/b)	0 (weak)
3 (4b, 1y)	−1 (electric)
4 (2b/y, 2y/b)	0 (weak)

Proton being pulled apart

Figure 3.60 *When bonded quarks are pulled apart, an internal electrical force of attraction builds within them, which tries to pull them back together. In the example shown, the quarks of a proton are being pulled apart. As this occurs, a region of negative electric charge develops in its interior, which interacts with its positive charge, driving the quarks back together. In the example, one unit of positive charge interacts with one unit of negative charge within the proton. However, this magnitude would increase the further the quarks were pulled apart, as more positive and negative regions would develop.*

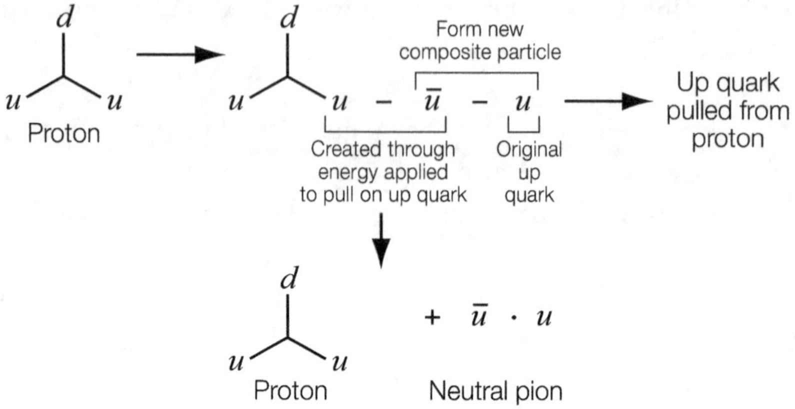

d

$u \qquad u$

Proton

$\longrightarrow$

d

$u \quad u \; - \; \bar{u} \; - \; u$

$\underbrace{\qquad}$ Created through energy applied to pull on up quark

Original up quark

Form new composite particle

$\longrightarrow$ Up quark pulled from proton

$\downarrow$

d

$u \qquad u$

Proton

$+ \; \bar{u} \cdot u$

Neutral pion

Figure 3.61 *The amount of energy needed to pull quarks apart is enough to create new quarks, such that there are always quarks attached to other quarks.*

There is a weak repulsive force within the neutron, owing to there being two weak charges of +2/3 associated with the down quarks. But similar to the scenario in which the quarks within a proton are pulled apart, an internal attractive force builds up within the neutron the further the quarks composing it are separated. It is just that in the case of the neutron, the attractive force holding it together under the stress of separation is associated with the weak force not the electric force. However, the effect is the same; the further the

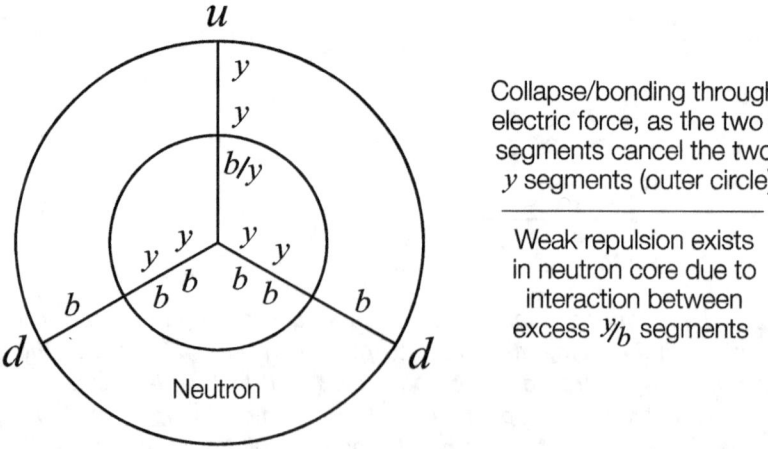

Collapse/bonding through electric force, as the two *b* segments cancel the two *y* segments (outer circle)

Weak repulsion exists in neutron core due to interaction between excess y/b segments

Figure 3.62 *The neutron (udd) is held together primarily through the electric force.*

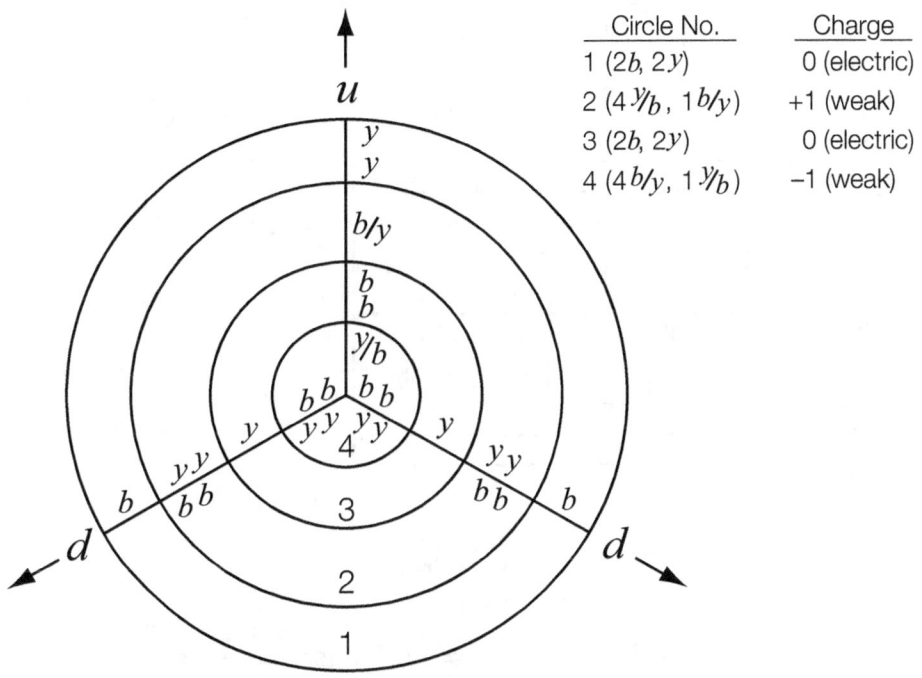

Circle No.	Charge
1 (2b, 2y)	0 (electric)
2 (4 y/b, 1b/y)	+1 (weak)
3 (2b, 2y)	0 (electric)
4 (4b/y, 1 y/b)	−1 (weak)

Neutron being pulled apart

Figure 3.63 *When the quarks of neutrons are pulled apart, an internal weak force of attraction builds within them, which tries to pull them back together.*

quarks are pulled part, the greater the magnitude of the attractive force pulling them back together (*Figure* 3.63).

Note that a proton is a well-balanced particle internally, having an equal number of bluish-green and yellowish-green segments at its core. The neutron has an excess of yellowish-green segments; see *Figure* 3.62. It is known to be slightly more massive than the proton. From the standpoint of TET, this is a result of those yellowish-green segments at its core, which add t⁺ to that area. As mass is t⁺ energy, this added t⁺ looks like mass within the particle.

Mesons are quark/anti-quark combinations. Two types are shown in *Figures* 3.64 and 3.65, the positive and negative pions, composed of an up quark and an anti-down quark and an anti-up quark and down quark, respectively. Note that any particle with three excess blue segments, such as an anti-proton and a

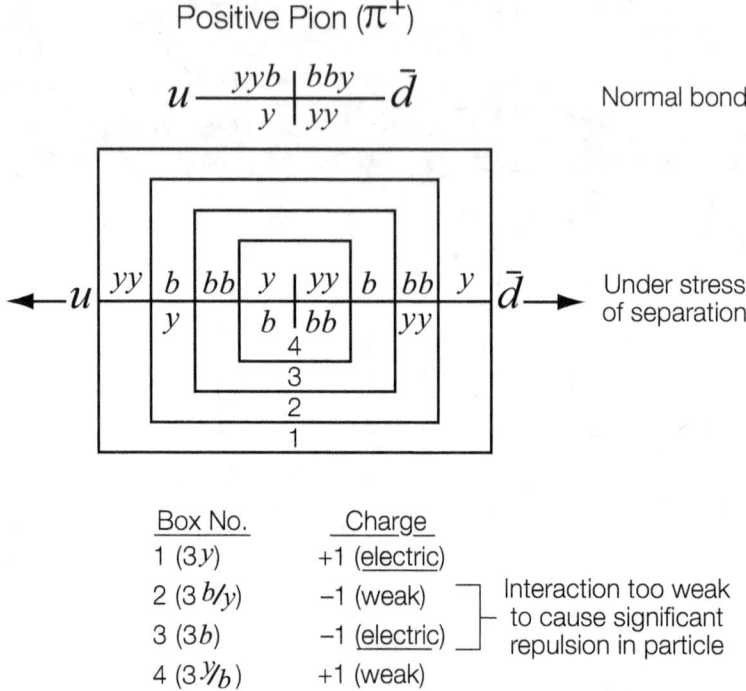

Positive Pion (π^+)

$$u \frac{yyb \mid bby}{y \mid yy} \bar{d}$$ Normal bond

$\leftarrow u$ ... $\bar{d} \rightarrow$ Under stress of separation

Box No.	Charge	
1 ($3y$)	+1 (electric)	
2 ($3\,b/y$)	−1 (weak)	Interaction too weak
3 ($3b$)	−1 (electric)	to cause significant
4 ($3\,y/b$)	+1 (weak)	repulsion in particle

Figure 3.64 *A positive pion is composed of an up quark and an anti-down quark.*

negative pion, will have one unit of negative charge equivalent to an electron. And likewise, any particle with three excess yellow segments, such as the proton and positive pion, will have one unit of positive charge equivalent to a positron. What is meant by *excess* is that there are no segments of opposite polarization in the particle frame to essentially cancel out the segments' effects. This is the case in the neutron, however, in which the electrically negative segments are balanced by electrically positive segments (*Figure* 3.66). Note that any excess weak charge within baryons and mesons are in a sense buried within the particles. All that is visible from the outside is the electric charge, or lack thereof, of their frames. The field lines emanating from electrically charged composite particles are the same as those generated by electrons and positrons. In the case of protons, anti-protons, and similar particles, however, this is only the case some distance from those particles. The areas of the field lines closest to the particles are a little more complicated. This will be further explored in the chapter 7.

Negative Pion (π^-)

$$\bar{u} \underset{b\ \ |\ bb}{\overset{bby\ |\ yyb}{\rule{3cm}{0.4pt}}} d$$ Normal bond

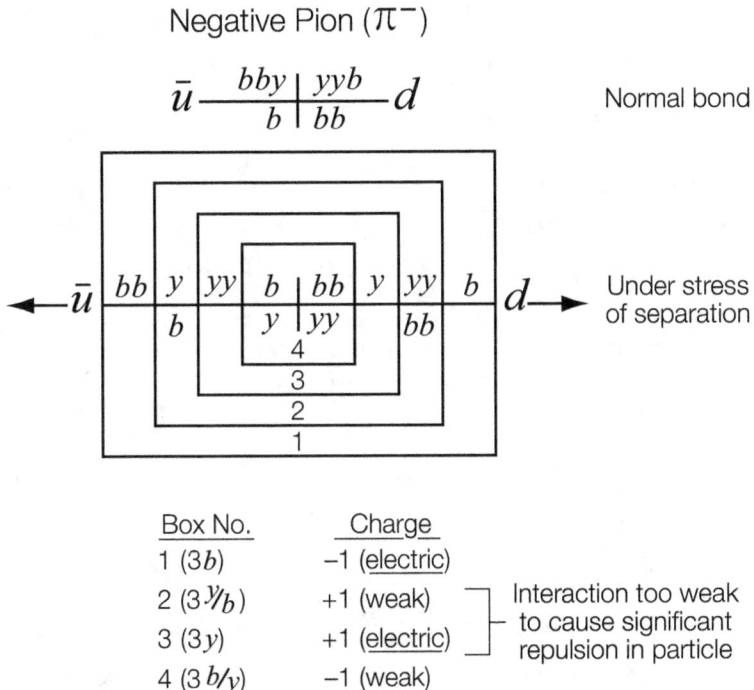

Under stress of separation

Box No.	Charge	
1 (3*b*)	−1 (electric)	
2 (3 $\mathcal{Y}_b$)	+1 (weak)	Interaction too weak
3 (3*y*)	+1 (electric)	to cause significant
4 (3 *b/y*)	−1 (weak)	repulsion in particle

Figure 3.65 *A negative pion is composed of an anti-up quark and a down quark.*

In QM, the strong force is mediated by particles called gluons. In TET, gluons are akin to photons and Z bosons. Whereas photons are composed of blue and yellow segments and temporal particles and Z bosons are composed of bluish-green and yellowish-green segments and temporal particles, gluons are composed of either blue and yellowish-green segments or yellow and bluish-green segments and temporal particles (*Figure 3.67*). In TET, a strong field is a static system of strong field lines, whereas a moving system of these lines is a strong-magnetic field. As electrons and positrons arise from an electromagnetic field and represent electrically charged photons, and electron neutrinos and anti–electron neutrinos arise from weak-magnetic fields and represent weakly charged Z bosons or W bosons, quarks and anti-quarks arise from strong-magnetic fields and represent strongly charged gluons in TET—although strong charge is really just electric charge (*Figure 3.68*).

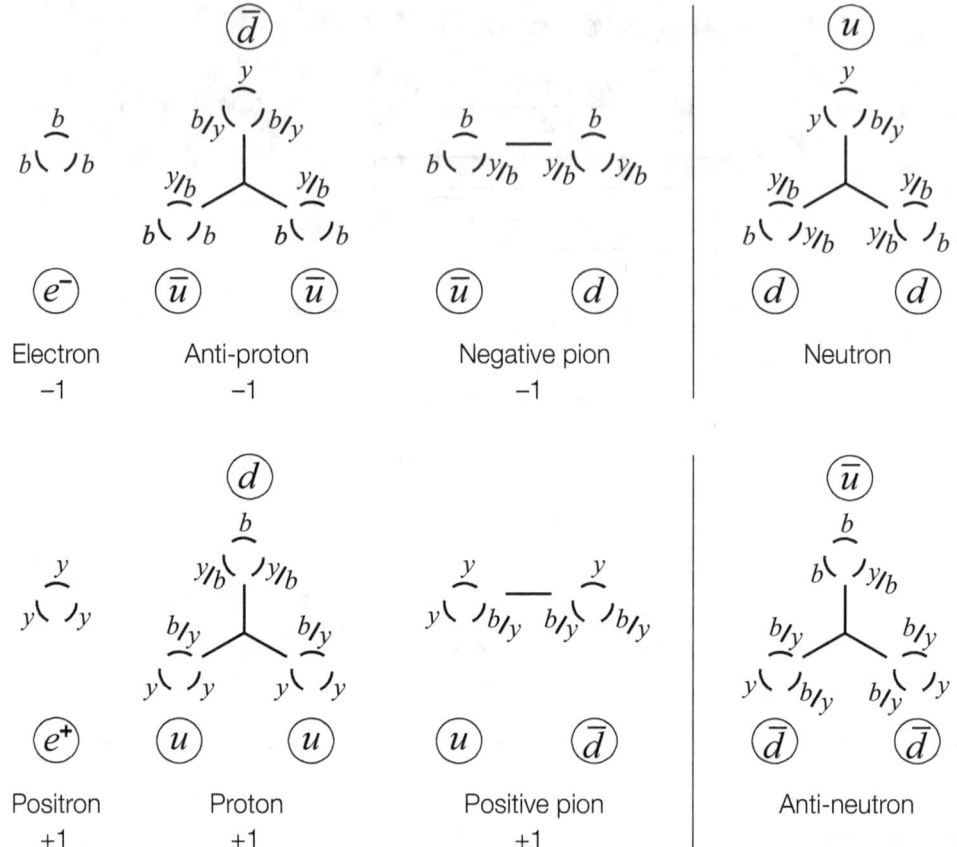

Figure 3.66 *Any particle with three excess blue segments will have one unit of negative charge similar to an electron. Any particle with three excess yellow segments will have one unit of positive charge similar to a positron. If the number of blue segments equals the number of yellow segments within a particle, it will be electrically neutral.*

$$\underline{\underline{bbb \mid yyy}} \longrightarrow \text{A photon}$$

$$\underline{\dfrac{bbb \mid yyy}{yyy \mid bbb}} \longrightarrow \text{A Z boson}$$

$$\underline{\dfrac{yyb \mid bby}{y \mid b}} \longrightarrow \text{A gluon}$$

Figure 3.67 *Gluons are similar to photons and Z bosons. They are composed of either yellow and bluish-green segments and temporal particles or blue and yellowish-green segments and temporal particles.*

Stationary quarks and strong field lines, along with their associated temporal particles, represent virtual gluons. Moving quarks and strong field lines, which develop a wave nature, represent real gluons (*Figure 3.69*). The blue and bluish-green segments of quarks, as well as the blue side of their yellowish-green segments, qualifies them as full magnets, like electrons or electron neutrinos, as

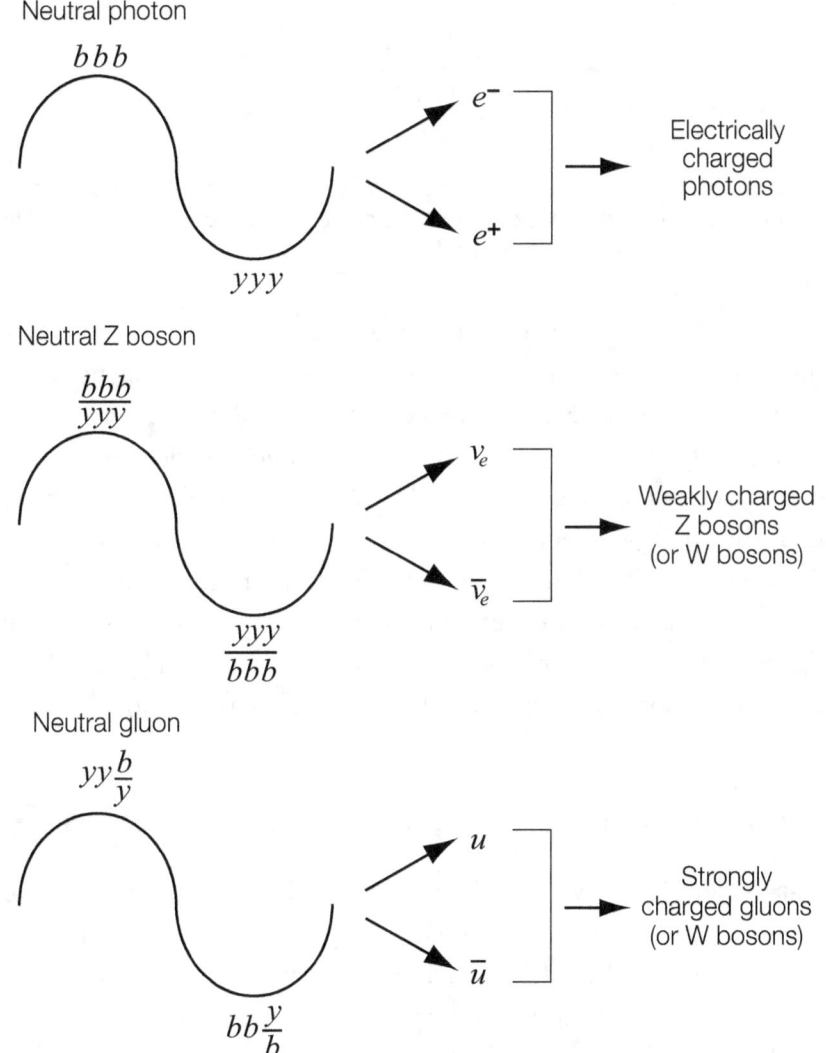

Figure 3.68 *Quarks and anti-quarks arise from gluons like electrons and positrons arise from photons and electron neutrinos and anti–electron neutrinos arise from Z bosons.*

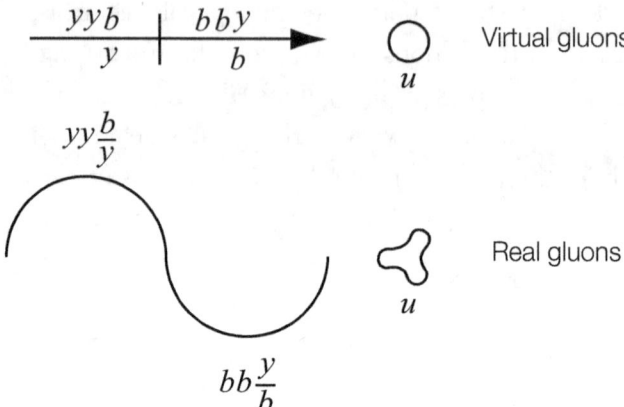

Figure 3.69 *Stationary quarks and strong field lines, along with their associated temporal particles, represent virtual gluons. Moving quarks and strong field lines, which develop a wave nature, represent real gluons.*

the blueness results in a loss of t^+ energy, a necessary phenomenon for full magnetism. Thus, as photons are involved in the electromagnetic force and Z and W bosons are involved in the weak-magnetic force, gluons are involved in the strong-magnetic force. (Note that, as mentioned previously, quarks are sometimes involved in the weak-magnetic force and as such also represent W bosons.) In TET, the strong-magnetic force is just a variation of the electromagnetic and weak-magnetic forces. Ultimately, both the weak-magnetic and strong-magnetic forces are derivatives of the electromagnetic force. As such, Z bosons and gluons can be considered different types of photons. The strong-magnetic force, like the weak-magnetic force is short ranged due to the weakly charged elements in the field lines.

Maintenance of Strong Field Lines

Similar to electric and weak field lines, the ultimate process maintaining strong field lines is the t^+/t^- process of the blue or bluish-green segments (*Figure 3.70*). As with the weak and electric field lines, the t^- particles that the blue and bluish-green segments give off are used by the yellow and yellowish-green segments to maintain their polarization. As with the other field lines, energy is continuously added to the system as t^+ and lost by the system as t^- (*Figure 3.71*).

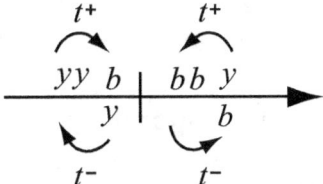

Figure 3.70 *As with the weak and electric field lines, the ultimate process maintaining strong field lines is the t^+/t^- process of the blue or bluish-green segments. The t^- particles that the blue and bluish-green segments give off are used by the yellow and yellowish-green segments to maintain their polarization.*

Maintenance of Strong-Magnetic Force

The maintenance of the strong-magnetic force occurs in a similar fashion to the maintenance of the electromagnetic and weak-magnetic forces. From the stand-point of TET, this means through 1) the t^-/t^+ process regarding the creation of virtual and real gluons and 2) the t^+/t^- process regarding the maintenance of strong field lines and frames. Again, however, although QM has processes equivalent to t^-/t^+ built into it to create virtual and real gluons, it has no process equivalent to t^+/t^-. QM regards the mere existence of the gluons as "good enough" for the maintenance of the force, thus focusing only on its energy-borrowing process (t^-/t^+), as this process supplies the energy to create the force carriers. As noted for the electromagnetic and weak-magnetic forces, the problem is that focusing only on the t^-/t^+ process leads to infinite energy results, as t^+ accumulates as time progresses but is not paid back as quickly. Thus, the

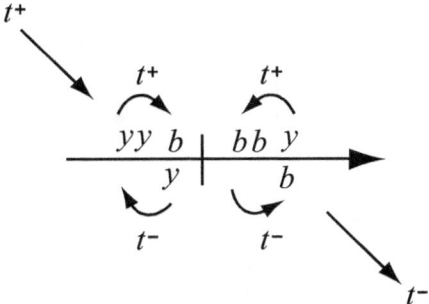

Figure 3.71 *As with the other field lines, strong field lines continuously borrow energy in the form of t^+ from exterior space and repay that energy in the form of t^-.*

ad hoc procedure of renormalization must be used to eliminate the infinite energy, which is equivalent to recognizing the t^+/t^- process as described earlier.

Quantum Mechanics and Temporal Energy Theory

The previous discussion of the electric, weak, strong, and magnetic forces by using TET concepts is not to suggest that there is anything wrong with QM. It is to suggest that perhaps QM is an approximation of TET. Many of the features of QM can be explained through TET concepts and in more detail, such as the existence of virtual particles and their role in bringing about the forces, the borrowing of energy from the vacuum to create the virtual particles, and the existence of a negative energy vacuum and a universal background clock. TET can also be used to eliminate in a more natural way the infinities that afflict QM, providing a possible physical basis to the mathematical procedure of renormalization.

However, the main takeaway message from this chapter regarding the potential relationship between QM and TET is that QM's description of the electric, weak, strong, and magnetic forces appears to rely on TET's t^-/t^+ temporal-conversion process, ignoring the t^+/t^- process. As discussed in the previous chapter, however, GR appears to rely on TET's t^+/t^- temporal-conversion process in describing gravity. Thus, when all of the forces are united into a single framework, the t^+/t^- process of gravity goes head to head with the t^-/t^+ process of the electric, weak, strong, and magnetic forces, leading ultimately to a zero result, which suggests there is no time in the universe.

CHAPTER
4

MERGING THEORIES

As stated in the Introduction, scientists have been trying to combine General Relativity (GR) and Quantum Mechanics (QM) into a single theory called the *Theory of Everything*. What they mean by "everything" is the total of the known fundamental forces of nature—gravity and the electric, weak, strong, and magnetic forces. The problem that arises when they combine all the mathematics is that an equation pops up, called the Wheeler-DeWitt equation, that says the universe has no time. They call this problem the *problem of time*. It is a problem because, not only do we intuitively know time exists, it plays an important role in both GR and QM, two theories that describe the workings of the universe well. Time must exist. But why then does the mathematics say it does not?

The Problem of Time

The root of the problem of time is currently viewed in the scientific arena from the perspective of the conflicting views GR and QM have regarding the conservation of energy with time. That is, QM allows increases in energy to a system that are associated with nothing more than simple time progression, such as the creation of a virtual particle cloud around a stationary charge, like an electron. The cloud simply arises from the vacuum and surrounds the electron as time progresses. Indeed, it is the progression of time that allows you to see mathematically the increases in energy. GR, however, disallows such changes in energy to a system—the system in that theory being the universe as a whole, with any energy changes coming from spacetime/the vacuum. That is, it disallows energy changes to the vacuum that are associated with nothing more than simple time progression. In GR, if energy were added to the vacuum with time, the added energy would vanish. If energy were taken away, the removed energy would be replaced. This maintenance of energy with time is ultimately enforced by the activities occurring at the level of white holes, from which energy is added to our vacuum, and black holes/singularities, into which energy disappears, and is the principal reason the vanishing of gravitational field energy during gravity has never caused a crisis in GR.

The combination of GR and QM (GR-QM) involves GR being placed into QM, a procedure sometimes referred to as the "quantization of GR." Typically, QM's external time parameter, its universal clock, would be used to solve the QM equations, but this time parameter is thought to have no meaning in GR, as there is no apparent universal clock in that theory. Therefore, some other aspect of QM must serve in the role of time when QM is combined with GR—some aspect of QM that GR can recognize. Because GR can recognize energy, the role of time in GR-QM is played not by the external time parameter itself, but by the increases in energy associated with that parameter. This is a legitimate procedure because in QM the two are linked. The connection between them is simply being viewed in reverse: In QM alone, the progression of time is used to see changes in energy, but in GR-QM, QM's changes in energy are used to see the progression of time. Note that with its steady time parameter on the sidelines, QM uses GR's steady expansion of the universe to help it see its changes in energy, which are then used to see the changes in time.

The problem is that GR blocks those changes in energy. It does this because they arise from the vacuum, and GR has running in the vacuum two opposing energy processes: one that adds energy (relating to white holes) and one that removes energy (relating to black holes/singularities). In GR, the energy that QM adds is indistinguishable from the energy added to the vacuum via white holes, such that any energy that QM adds will be matched by a loss of energy at singularities, as the energy of white holes is in general. As a result, the added energy never stands out—there never appears to be a change in the energy level of the vacuum. As QM's changes in energy were being used to see changes in time in GR-QM, the universe as viewed from the vantage point of GR-QM appears timeless. Put another way, GR and QM simply appear incompatible: One allows changes in energy with changes in time only (or vice versa), and the other does not. When they are combined, QM becomes affected by the energy-change constraints GR places on itself, such that QM's energy changes are annihilated. And as those energy changes were defining time progression in the GR-QM framework, the universe, as viewed from the GR-QM perspective seems to be stuck in a single, unchanging moment.

From the perspective of TET, the energy-change constraints GR places on itself, as well as GR-QM, ultimately relate to how the positive and negative energy vacuums are treated in that theory. In GR, the positive energy vacuum, in which the t^+/t^- process dominates, and the negative energy vacuum, in which the t^-/t^+ process dominates are treated equally, even though GR's focus is on the positive vacuum. The equal treatment that GR gives to the vacuums is evidenced by the equal importance it gives to singularities, at which level t^+/t^- dominates, and white holes, at which level t^-/t^+ dominates. White holes are simply what the negative vacuum looks like from the perspective of the positive vacuum.

By giving equal importance to the two vacuums, GR also gives equal importance to their time processes, such that GR incorporates two codominant but oppositely running time processes at once. As a result, the energetic events associated with one time process cancel the energetic events associated with the other time process, such that neither is allowed to upset the balance of energy in the universe. For example, from the perspective of the positive vacuum, as time progresses in the t^+/t^- direction at the level of singularities, t^+ energy is lost, but as time progresses in the t^-/t^+ direction at the level of white holes, t^+ energy is gained. As the two time processes occur simultaneously, there is no net change in energy with time progression in the universe. If GR also focused on the negative vacuum, it would see the loss and gain of energy in the form of t^- in that vacuum as time progressed, with no net change; with GR focusing on the positive vacuum only, the events involving t^- occur, in a sense, "under the surface," out of view (*Figure 4.1*).

Note that t^+/t^- and t^-/t^+ each represent *two* fundamental moments of time. For t^+/t^-, they are t^+ to t^0 and t^0 to t^-. For t^-/t^+, they are t^- to t^0 and t^0 to t^+. If broken down into fundamental moments, the positive vacuum gains energy

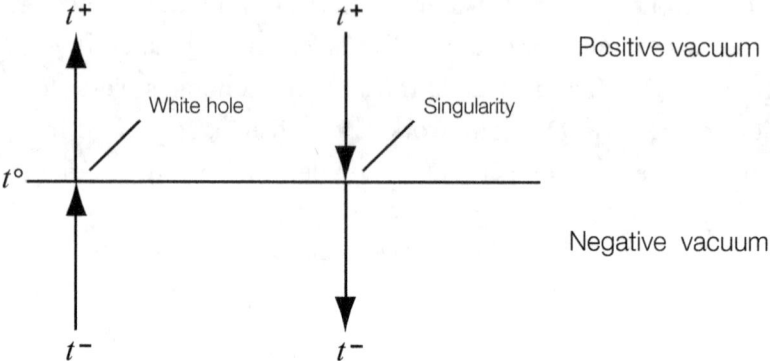

Figure 4.1 *If GR also focused on the negative vacuum as it does the positive vacuum, it would see the loss and gain of negative energy (t⁻) in that vacuum as time progressed, with no net change.*

over a single, fundamental, moment of time (t^0/t^+) at the level of white holes and simultaneously loses energy over a single fundamental moment of time (t^+/t^0) at the level of matter/singularities. (In the negative vacuum, energy is gained during t^0/t^- and lost during t^-/t^0.) The importance of this is that energy is gained and lost over moments of time that are not further divisible. They are time at its foundation, and indeed its fastest. Thus, the level of energy never appears to increase or decrease with the progression of time.

From the perspective of TET, QM does not have the energy-change constraints GR does—that is, it allows changes in energy to a system with simple time progression—because it does not treat the positive and negative energy vacuums equally, although it officially recognizes both. Specifically, and again from the TET perspective, QM sends energy from the negative vacuum to the positive vacuum faster than it returns it. That is, QM ties the borrowing of energy by a system in the positive vacuum, such as a field of virtual photons, to the negative vacuum's t^-/t^+ time, with the t^+ the negative vacuum produces being the energy that is borrowed by the system. However, QM does not tie the payback of that energy to the system's t^+/t^- time. There is no process inherent to QM that is equivalent to the t^+/t^- process at all. Energy is paid back but on a much slower basis, giving the sense of it lingering in the positive vacuum, if only for a small interval of time. As time progresses—ultimately in the t^-/t^+ direction, as this is the only time QM appears to recognize—there is

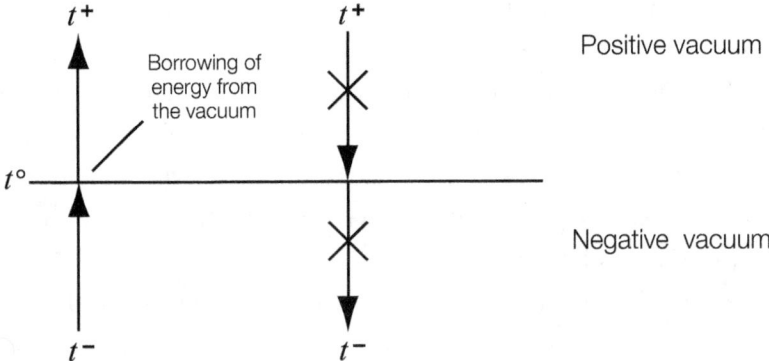

Figure 4.2 *In QM, energy borrowed from the vacuum is paid back quickly, but not via the fundamental progression of time. Therefore, as time progresses, there is a discernible, mathematical increase in energy.*

a discernible, mathematical increase in the energy level of the system, because there is no t^+/t^- process to balance the other process (*Figure 4.2*).

Note that because t^-/t^+ stems from the negative vacuum, it behaves as an external time parameter from the standpoint of the positive vacuum (our space)—that is, the negative vacuum and thus its time are external to the positive vacuum. As noted earlier, when GR and QM are combined, time in the GR-QM framework is defined not by the external time parameter, but by the energetic events associated with it, events akin to the creation of a virtual particle cloud around a stationary charge. Thus, from the standpoint of TET, it is the t^+ energy added to the positive vacuum through the negative vacuum's t^-/t^+ process that defines time in the GR-QM framework. However, this added t^+ energy is consumed by GR's t^+/t^- process. That is, through this process, the energy added in the form of t^+ is removed in the form of t^-. It is made to vanish. Thus, time itself appears to vanish, leading to the problem of time.

Note that defining time by the increase in t^+ energy in the positive vacuum is equivalent to defining time by the negative vacuum's t^-/t^+ process itself—the external time parameter—as it is that process that causes the increase in t^+. This means that, ultimately, viewing changes in energy to see changes in time is exactly the same thing as viewing changes in time to see changes in energy. It is a "distinction without a difference," because the energy involved is temporal energy. Both views refer to the t^-/t^+ process from the TET perspective. Thus, the more complete answer as to why there is a problem of time in GR-QM, from the standpoint of TET, is that the t^+/t^- process of GR is going head to head with the t^-/t^+ process of QM, such that they cancel each other out. From this perspective, QM's t^-/t^+ process overlaps with the t^-/t^+ process occurring at the level of white holes in GR. Thus, like the flux of energy at white holes, QM's flux becomes obscured, as it is countered by GR's t^+/t^- flux. The increases in t^+ energy in the positive vacuum brought about by QM's t^-/t^+ process are essentially washed away by the opposing time process in GR. With this cancellation, the past, present, and future look exactly the same mathematically. Again, the universe appears to be stuck in a single, unchanging moment in time (*Figure 4.3*).

In a certain way, the problem of time looks like a problem principally with GR, in that GR itself contains two equal, yet opposite time processes that cancel each other out mathematically. However, white holes along with their backward time process are considered to be mere mathematical curiosities. They

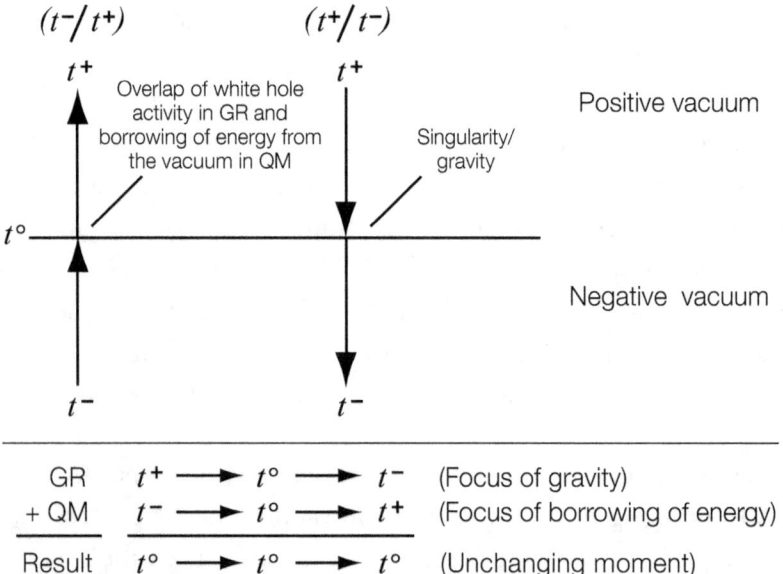

Figure 4.3 *QM's energy flux (t^-/t^+) is countered by GR's gravitational energy flux (t^+/t^-), as the energy flux of GR's white holes is in general. The result is that the universe appears to be stuck in one unchanging moment of time.*

are not considered to have physical existence. Thus, GR is not considered to have a problem of time in itself. Interestingly, if the negative energy vacuum of QM were treated similarly to the positive vacuum, such that energy is paid back to the negative vacuum as quickly as it is borrowed, QM itself would be said to have a problem of time. This is because it would be recognizing two equal, yet opposite time processes, whose energetic events cancel each other. As t^-/t^+ would add t^+ to the positive vacuum (removing t^- from the negative vacuum in the process), t^+/t^- would remove t^+ from the positive vacuum (adding t^- to the negative vacuum in the process). As the negative vacuum is the source of the t^-/t^+ process, the positive vacuum would be the source of the t^+/t^- process.

The recognition of the t^+/t^- process occurring in the positive vacuum in QM is equivalent to regarding QM's renormalization process, not as a mathematical trick, but as representative of a real physical process. In renormalization, excess positive energy is cancelled out by negative energy, which is equivalent to the transformation of t^+ to t^- and the disappearance of that t^- from the positive vacuum as it is absorbed by the negative vacuum (see *Figure 3.44*). As a mathematical trick, renormalization poses no problem of time in QM, but as a true physical process it would, because it would mean that time is truly running in two directions in QM, canceling each other out and, overall, producing a zero time result.

When GR and QM are combined, GR's gravity takes the role of renormalization, but as a true physical process with forward-in-time motion, and QM's borrowing of energy from the vacuum takes the role of white hole activity, but also as a true physical process with backward-in-time motion. What this does is simply make the problem of time unavoidable. Renormalization can be regarded as just a mathematical trick, but gravity is a true physical process that cannot be ignored. White holes can be regarded as just a mathematical curiosity, but the borrowing of energy from the vacuum is a true physical process that cannot be ignored. As gravity involves motion forward in time (t^+/t^-) and the borrowing of energy from the vacuum involves motion backward in time (t^-/t^+), the two time processes cancel each other out, and the universe appears timeless.

Put yet another and final way, GR is principally a theory of gravity, and as such, its primary focus is on the t^+/t^- process, as this is the process that makes gravitational field energy vanish, producing a full gravitational attraction between two objects. QM is principally a theory of the electric, weak, strong, and magnetic forces, and as such, its primary focus is on the t^-/t^+ process, as this is the process that creates the virtual and indeed real particles that produce the forces—t^+ is essentially the meat of those particles. White holes are a secondary focus in GR, having no direct role in gravity. Renormalization is a secondary focus in QM, having no direct role in the electric, weak, strong, and magnetic forces. In GR alone, the secondary focus of white holes can largely be ignored. In QM alone, the secondary focus of renormalization can be considered non-physical. However, when GR and QM are combined, the two primary foci, which cannot be ignored as they represent the forces, go head to head with each other and cancel each other out, and again, the universe appears timeless.

GR's Hidden Universal Clock

Note that in the current formulation of QM, and as is ostensibly the case in GR, the negative vacuum is hidden behind the positive vacuum. It is just that in QM, this is done deliberately through another renormalization step. As the negative vacuum is the origin of the external time parameter from the standpoint of TET, it is understandable, from the perspective of the positive vacuum, for the source of this parameter to seem like some mysterious, separate thing that ticks away regularly "somewhere out there" in the background, because the negative vacuum is indeed separate from the positive vacuum and thus its time does exist outside of the positive vacuum.

In GR, this time parameter, as an external phenomenon, is largely invisible because GR focuses on the positive vacuum only, and the t^-/t^+ process stems from the negative vacuum, which is "under the surface" of the positive vacuum, out of view in that theory. The t^-/t^+ activity at white holes is not considered to be external because it is viewed from the standpoint of the positive vacuum. Also, the external time parameter is considered to be a single, overarching time in the universe, but there are many white holes, which presumably are their own clocks. Similarly, every matter, anti-matter, and force carrier particle is its own clock. Thus, because GR focuses on the positive vacuum, which houses many such particles, as well as many white holes, it is considered to have many clocks, none of which can be said to have the "correct" or overall time. The invisibility, or apparent absence, of the external time parameter and the presence of many clocks, each with its own time, make it seem as though the external time parameter has no meaning and is indeed unusable in GR.

The Non-Problem of Time

The zero time result that comes from the combination of GR and QM is not really a problem. One of the principal reasons is that it has no direct bearing on thermodynamic time, in which neutral photons travel away from their source particles never to spontaneously return, giving time a definite direction. Thus, thermodynamic time is cause-and-effect time. This type of time is the reason why faded paint does not spontaneously become bright again, a melted ice cube does not spontaneously refreeze, or a cracked egg does not re-form out of the blue. It is related to increased disorder, or entropy, in the universe.

Composite structures—composites of elementary matter or anti-matter particles held together by force carriers—experience thermodynamic time. The elementary matter, anti-matter, and force carrying particles themselves only experience t^+/t^- and/or t^-/t^+ time, each of which represents basic time. This form of time has no direct bearing on thermodynamic time. The influence of the t^+/t^- and t^-/t^+ processes on thermodynamic time is only indirect, supportive. That is, both processes are involved with the maintenance of an electromagnetic field, for instance, but this field will still carry energy, in the form of photons of visible light or heat, for example, away from its source never to spontaneously return, providing a definite thermodynamic direction to time.

Another reason the problem of time is really a non-problem is that when GR and QM are united, they become involved in a "you scratch my back; I'll scratch yours" dynamic, in which GR's gravity takes the role of QM's renormalization process to eliminate QM's infinite energy problem, and QM's borrowing of energy from the vacuum takes the role of GR's white holes to replenish GR's lost gravitational field energy. Thus, the two theories are not incompatible;

they are complementary. They are opposite sides of the same energy coin and the same temporal coin, as those energy changes are connected to opposing temporal processes.

Another way of describing their complementary nature is that QM needs GR's white holes to borrow energy from the vacuum and GR's gravitational process/ singularities to get rid of the energy once it is no longer needed. GR needs the discrete manner in which QM borrows and returns energy to see the actual flow of energy at its white holes and singularities, which helps it also see the more granular nature of the gravitational field, as opposed to the erroneous view of the field as something smooth and continuous. It has been known for some time that the gravitational field cannot be smooth and continuous at very small scales, which was a clue to its overall particulate nature. Thus, the problem of time is really a non-problem. The zero time solution that results from the combination of GR and QM is the natural, correct answer, and indeed, the two theories actually need each other to overcome their individual shortcomings.

The so-called problem of time is really just one obstacle to combining GR and QM into a single framework. The main goal in joining these theories is show- ing how gravity and the electromagnetic, weak, and strong forces are just dif- ferent manifestations of a single force. At this point in history, attempts to unify these forces fall into two steps. The first is finding a theory that unifies the electromagnetic, weak, and strong forces—such a theory has been called the Grand Unification Theory. In this framework, these three forces are thought to be different manifestations of a single force that can be called the Grand Unified force. The second is finding a theory unifying the Grand Unified force with gravity—this has been called the Theory of Everything. The sections that follow describe TET's perspective on Grand Unification and the Theory of Everything.

The Grand Unification Theory

Grand Unification has really already been discussed in the previous chapter, though the term is introduced in this chapter. In TET, the Grand Unified force is simply the electromagnetic force, in that the weak and strong forces (or weak-magnetic and strong-magnetic forces in TET) are just different manifestations of the electromagnetic force. As noted earlier, the connection between the electromagnetic and weak forces has been known for some time, such that together they have been called the electroweak force. The weak/weak-magnetic force is just a weak version of the electromagnetic force from the TET perspective. And the strong/strong-magnetic force is also related to the electromagnetic force, having both weak and electromagnetic characteristics.

The Theory of Everything

The Theory of Everything is again the theory that unites the electromagnetic, weak, strong, and gravitational forces. As the weak and strong forces are just different manifestations of the electromagnetic force, the Theory of Everything ultimately reduces to the union between the electromagnetic and gravitational forces. For the purposes of this discussion, however, it is helpful to separate electromagnetism into its electric and magnetic components and thus regard the Theory of Everything as the union of the electric, magnetic, and gravitational forces. The question is, What is the single force at the heart of these three forces?

The answer in TET is the electric force. Magnetism and gravity in their fullest senses rely on the t^+/t^- process. That is, for these forces to fully occur, t^+ must be converted to t^-, which is subsequently drained away. Although there could be some gravitational and magnetic attraction without t^+/t^- (due to the mutual attraction between temporal particles), these forces would be muted without the loss of energy from the systems brought about through t^+/t^-. The ultimate elements capable of performing the t^+/t^- process are blue (electric) elements in field lines and particle frames. Indeed, even yellow elements in field lines and frames depend on blue elements, as described earlier. Bluish-green elements are also capable of processing time in the t^+/t^- direction, but they are simply weakly electric.

Note also that the minimal magnetism and gravity that yellowish-green and true-green elements are able to bring about comes from their blue "sides." As discussed in chapter 5, positrons, the only QM particles completely incapable of processing time in the t^+/t^- direction, being completely yellow, are unable to actively participate in magnetism or gravity. Thus, both the gravitational and magnetic forces rely on the electric force, particularly the blue, forward in time side of the electric force. Without the electric force, gravity and magnetism as we know them would not occur. Indeed, the electric force reigns supreme over all the other forces. A watered-down version of the electric force represents the weak force, and the weak force combined with the full electric force represents the strong force. Finally, the conversion of t^+ energy to t^- energy by blue and blue-related elements allows full gravity and magnetism to occur.

Under normal conditions the five forces—electric, weak, strong, magnetic, and gravitational—look very different, but under high-energy conditions, they would be indistinguishable. In physics, small space is equivalent to high energy, so the inability to distinguish the forces on a small scale equates to an inability to distinguish them at high energy. Ignoring the fluctuations of interior space, during which t^+ and t^- are bounced back and forth, if you could see a temporal particle interact with a small region of space and then vanish, the following questions could be asked: (1) Was that the interaction of a temporal particle with an electric field, weak field, or strong field? (2) Did it represent the magnetic force, or (3) did it represent the gravitational force? Because the particle vanished, you would know that it at least involved the electric force, but it could easily have involved the weak, strong, magnetic, or gravitational force, as well. The main idea is that on that level, each of the five forces looks exactly the same. Indeed, they are the same force, the electric force, as this is the common denominator among them. Thus, in essence, the gravitational, magnetic, strong, and weak forces can be viewed as different manifestations of the electric force, making it the force at the heart of the Theory of Everything.

TEMPORAL ENERGY THEORY AND ADDITIONAL SCIENTIFIC CONCEPTS

CHAPTER
5

GENERAL RELATIVITY REVISITED

T his chapter revisits and introduces new General Relativity (GR)–related ideas, discussing them from the TET perspective. As noted in chapter 2, unlike GR, which considers gravity to be caused by the curvature of the so-called time dimension of spacetime, TET considers gravity to be caused by the flow of temporal particles toward matter and ultimately back into space. This of course involves the temporal respiration of matter, the conversion of t^+ to t^-, and the subsequent absorption of t^- by space. The volume of temporal particles (time) and the space they converge on are simply curved in the process of their moving toward the matter. Previously, it was stated that gravity is caused by the following factors in TET:

1. The attraction of t^+ particles to particles of matter (or energy generally);

2. The bonds t^+ particles form with each other and with a matter particle's mass—the more compact the temporal particles are, the stronger the bonds between them and also the mass of the matter particle they are connected to;

3. The conversion of t^+ to t^- by matter;

4. The absorption of t^- by exterior space.

In light of the past discussions, this list can be revised, such that in TET gravity can be said to be caused by:

1. The attraction of t^+ particles to energy;

2. The bonds the t^+ particles form with each other and with a QM particle's mass, which may be composed of t^+ and/or t^- particles—the more compact

temporal particles are, the stronger the bonds between them;

3. The conversion of t^+ to t^- by the QM particle—which occurs in all particles except the positron;

4. The absorption of t^- by exterior space.

The revision accomplishes several things: It focuses on the attraction of t^+ particles to energy generally rather than just to matter, which is more precise. By using the more collective term of "QM particle," it includes force carriers, as well as anti-matter, which except for the positron, also processes time in the t^+/t^- direction. Note, however, as revised item 2 indicates, that the t^+ particles involved in gravity would bond to the positron's mass, which is composed of t^- particles only (negative mass), as they bond to the mass of any other QM particle. This is because temporal particles bond with other temporal particles regardless of their spatiotemporal charge. All QM particles except the positron experience active gravity; the positron experiences passive gravity. That is, all QM particles except the positron actively participate in the gravitational experience through their t^+/t^- processes. The positron can be gravitationally pulled by another QM particle, but because it experiences only t^-/t^+ time, it cannot fully pull back, as such pulling involves the consumption of gravitational field energy. The positron's gravitational field simply hovers around the positron's mass. The positron also cannot actively participate in magnetic attraction and repulsion for the same reasons, as full magnetic attraction and repulsion also involves the consumption of t^+ energy.

The electron neutrino, muon neutrino, and tauon neutrino like most other QM particles actively participate in gravity. This is also likely true of their anti-neutrino counterparts, even though these particles predominantly process time in the t^-/t^+ direction. This is because, being largely green, they have a substantial blue nature causing t^+/t^-, in addition to their yellow nature causing t^-/t^+, and exterior space constantly seizes t^- particles. Thus, they will always be losing some t^-. However, because on average they are producing t^+ rather than eliminating it, the gravitational attraction they cause will likely move very slowly. Also, because of their slight preference for t^-/t^+ time, their gravitational field, as well as that of the positron, will likely swell. That is, positrons and anti-neutrinos should have stronger gravitational fields, in terms of the amount of t^+ surrounding them, than their matter counterparts, because positrons do not eliminate t^+ at all and anti-neutrinos eliminate it very little. Although not official QM particles, all three sterile neutrinos (electron type, muon type, and tauon type), which are true green all over, also likely participate in gravity actively, but slowly. And their gravitational fields also likely swell. In the particles, both the t^+/t^- and t^-/t^+ processes occur equally; thus like the anti-neutrinos, they are at times adding to their gravitational fields (through the addition of t^+) rather than subtracting from it.

Note that the mass of a QM particle is the volume of temporal particles responding specifically to the energy of its frame, with some of those particles directly bound to the frame. The gravitational field is the volume of temporal particles associated with the mass. Thus, gravity occurs through the mass field: A t^+ particle within the mass field is converted to t^0 then to t^-. The t^- is ejected into the gravitational field where it is absorbed by exterior space, and a new t^+ particle, ultimately from exterior space, takes the place of the departing t^+ (now t^-) particle and becomes part of the mass.

Any region of space with a length, width, and height of about 10^{-35} meters—the approximate room needed for a temporal particle—has the capacity to act as either a singularity (through which t^- exits interior space to enter exterior space) or a white hole (through which t^+ exits exterior space to enter interior space). What it will be depends on how much energy is in the area. Generally, singularities exist where energy is high; white holes exist where energy is low. Where energy is high, space is also curved due to t^+ particles compressing it as they move toward the source of that energy. Space is generally flat where energy is low. Thus, t^+/t^- is generally associated with curved space and of course gravity, and t^-/t^+ is generally associated with flat space and the injection of energy into our vacuum. Because interior space (and of course exterior space) is expanding, there are always new regions where t^+ particles can be emitted freely.

Some areas of interior space behave like both singularities and white holes, because interior space, being true green, processes time in both directions. This likely occurs in areas with a low to moderate matter content. In these areas, t^+ and t^- largely bounce back and forth—that is, sometimes t^+ exists and sometimes t^- exists, irrespective of any energy being absorbed at actual singularities or emitted at actual white holes. Time in these areas of interior space is ill defined, as sometimes it is moving forward (t^+/t^-) and sometimes backward (t^-/t^+); there is a great deal of uncertainty. The bouncing of t^+ and t^- back and forth within interior space is likely the quantum fluctuations of space in QM. That is, QM scientists have known for some time that volumes of space about 10^{-35} meters in length, width, and height jitter wildly with activity. In TET, this activity is the constant transformation of t^+ to t^- and t^- to t^+ by interior space, and thus any energy associated with this activity is the energy associated with the transformations of the temporal particles.

What Is Relative?

All of the structures of the universe are composed of just space, time, and the energy associated with these two elements. For example, the frame of an electron is just space, albeit temporally polarized space. When temporal particles congregate around a matter particle, for instance, they not only compress the space surrounding the particle, they compress the space that is the particle. That is, they compress the matter particle's frame. The state of compression of a particle's frame will affect not only its size, but also its time and mass.

Consider a large, composite matter particle. For demonstration ease, consider it to magically appear in space and for the elementary particles composing it to be arranged in spherical layers, or circular layers when viewing them in cross-section. As t^+ particles converge on this large particle, they will accumulate more within the center regions, where there is the greatest concentration of energy, with an increasingly more sparse concentration of them toward the outer edges. What this means is that more spatial compression will develop deeper inside the composite structure than toward its surface, which in turn means that the frames of the individual particles within the composite particle will be more compressed in the inner regions than in the

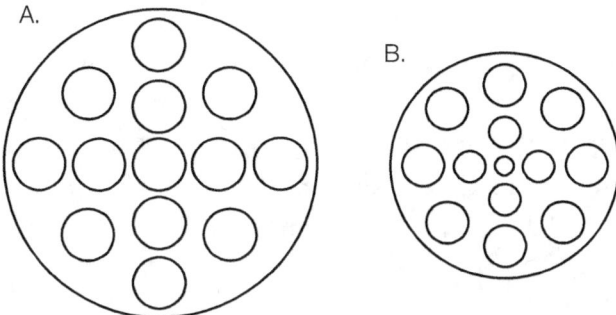

Figure 5.1 *When temporal particles congregate around a matter particle, for instance, they not only compress the space surrounding the particle, they compress the matter particle's frame. "A" represents an expanded composite particle. "B" represents the same particle but compressed due to the influx of temporal particles. Note that there is more compression toward the interior, where the concentration of energy is greater, than toward the outer edges.*

outer regions of that structure. Of course, the composite particle as a whole is also compressed when temporal particles converge on it (*Figure 5.1*).

When a QM particle—be it matter, anti-matter, or a force-carrier—is compressed, there is less surface area upon its frame to convert t^+ to t^-, or t^- to t^+. For example, consider two electrons, one in an expanded state and one in a compressed state (*Figure 5.2*). Each of the three segments composing the

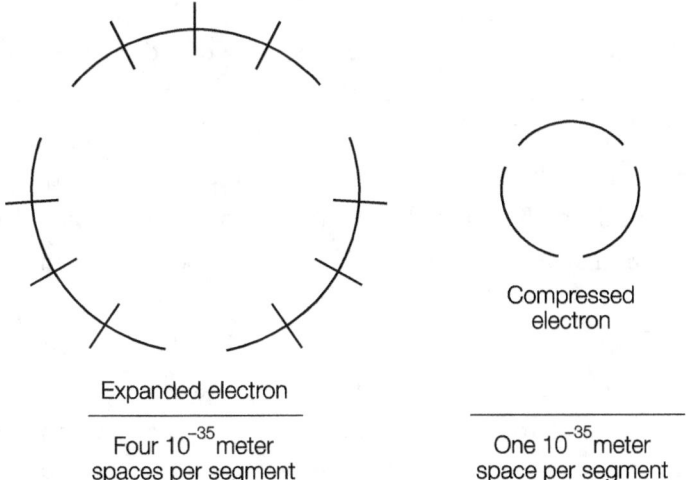

Expanded electron

Four 10^{-35} meter
spaces per segment

Compressed
electron

One 10^{-35} meter
space per segment

Figure 5.2 *When particles are compressed, they have fewer temporal spaces along their frames to process time.*

expanded electron's frame has several areas along its length that are 10^{-35} meters long, called temporal spaces, that can transform t^+ to t^-. The three segments of the compressed electron, because they are compressed, have fewer of these spaces. The result of this is that not only is the compressed electron's size smaller, its time is slower. Also, as described below, its mass is greater. Regarding the expanded electron, its size is larger. Its time is faster, and its mass is less.

The size difference between the electrons is very understandable, as one is compressed and the other is expanded. Anything composed of matter or anti-matter (and their force carrier particles) is capable of being compressed by temporal particles—from the tiniest speck of dust to our bodies to the planets and stars. Note, as this discussion has suggested, that the compression, for example of an apple, is not due to its atoms and molecules being squashed together. Instead, it is due to the shrinking of the frames of all the elementary particles within those atoms and molecules. That is, the apple's electrons shrink. Its quarks shrink. And the force carriers holding them together shrink. Because of this, the apple as a whole shrinks, but it is the same apple. Compression by temporal particles does not destroy an object's structural, functional, or chemical characteristics. If it is a crisp apple when expanded, it is a crisp apple when compressed.

The time and mass differences between expanded and compressed particles have to do with the number of temporal spaces along their frames. The time of a particle such as an electron corresponds to all of the t^+/t^- cycles occurring along its frame. When there are fewer temporal spaces, there are fewer t^+/t^- cycles, and thus the particle's time slows down. Note also that the number of temporal spaces corresponds to the number of field lines emanating from the particle. Thus, with a reduction in the number of spaces, there is a reduction in the number of field lines (*Figure* 5.3). The reduction in field line number, in turn, results in a reduction in thermodynamic time. Simply put, fewer field lines equate to fewer thermodynamic events. With fewer thermodynamic events, processes related to thermodynamics, such as aging, slow down. Thus, although basic (t^+/t^- and/or t^-/t^+) time has no direct bearing on

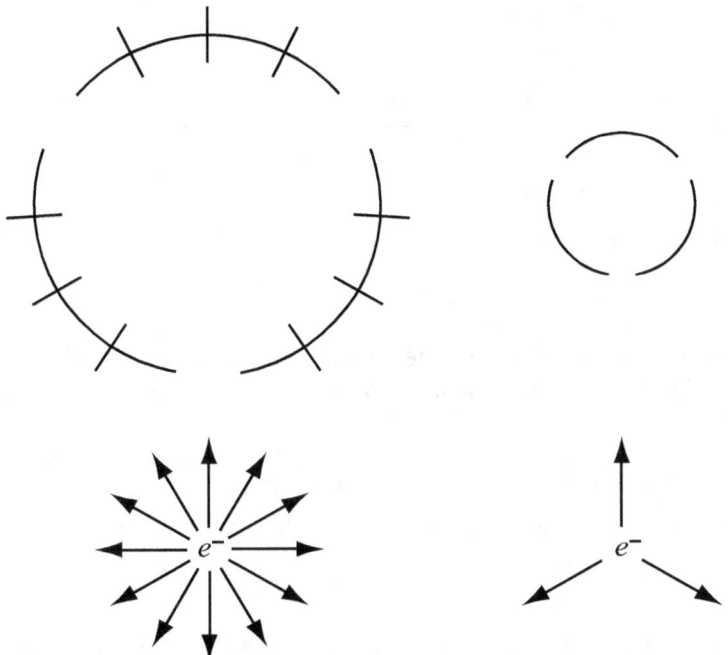

Figure 5.3 *When particles are compressed, they have fewer field lines emanating from them. The number of field lines corresponds to the number of temporal spaces in the frame.*

thermodynamic time, the two are correlated. If basic time slows, thermodynamic time slows as well.

Mass increases with fewer temporal spaces because the electron is able to carry more of its mass energy instead of processing it. That is, the expanded and compressed electrons attract the same number of temporal particles to their frames. The expanded electron, however, processes and thus gets rid of a greater amount of the temporal particles in its mass field as time progresses because it has more spaces by which to do so. The compressed electron, which has fewer spaces, is less able to do this and thus carries more of its mass energy (*Figure* 5.4). Also, temporal particles are more energetic when they are tightly packed, adding to the greater apparent mass of the compressed electron. The strength of the gravitational field around a compressed particle is greater because the temporal particles within the field, like those in the mass field, linger longer, becoming more tightly packed and thus increasing their energy

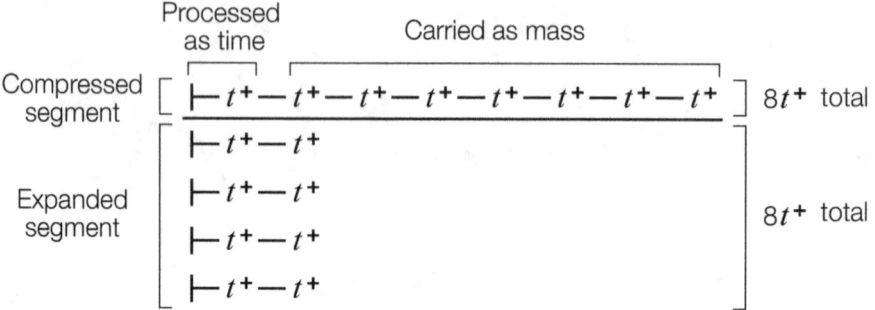

Figure 5.4 *Mass increases with fewer temporal spaces because the electron is able to carry more of its mass energy instead of processing it as time.*

and interaction strength, and of course, more continually move in. Thus, with fewer temporal spaces, not only is time slower and mass greater, but gravity is stronger.

When it comes to gravitational issues, the answer to the question—What is relative?—is thus size, time, and mass. The values of these factors are relative to the positions of particles within a gravitational field. Using the compressed composite particle from *Figure 5.1* and personifying the individual particles within it for a moment, it is interesting that each individual particle would perceive nothing out of the ordinary about its size, time, and mass, regardless of where it is in the gravitational field. Even if one were to move, for example, from the outer edge to deep inside the structure, it would notice nothing different. It would only perceive differences in the particles surrounding it if they are within a different density of temporal particles than it is, which the phrase "different reference frame" refers to in TET. That is, different densities of temporal particles represent different frames of reference. And there is of course no such thing as the "correct density." The densities simply are what they are, and thus, there are no universal correct values for size, time, and mass. They depend on the positions of particles within a gravitational field.

Recall that there is really only one gravitational field that extends throughout all of space—the gravitational field proper. Thus, even a particle far outside the composite structure cannot be said to have the correct size, time, and mass. Again, it depends on its position within the gravitational field, in this case the greater field extending throughout all of interior space. Objects within the same density of temporal particles look normal to each other. Objects in different densities appear to have different size, time, and mass values. These phenomena extend to the macroscopic level as well, although extra-sensitive equipment would be needed to perceive them. To the people at the top of a very tall office building, for example, the people at the bottom are smaller (literally, not just because they look smaller being so far away). And their mass is greater, and their time slower. To the people at the bottom of the building, the people at the top are larger. And their mass is less and their time is faster. Those at the bottom of the building are deeper within Earth's gravitational field. The people at the top of the building look normal to each other, and the people at the bottom of the building look normal to each other as well. To astronauts hovering far above the Earth, our size is smaller. Our mass is greater, and our time is slower. We are deeper within Earth's gravitational field than they are. We would not notice anything different about ourselves. To us, the astronauts are larger in size. Their time is faster, and their mass is less. However, they notice nothing out of the ordinary about themselves.

The Speed of Gravity

In GR, gravity occurs at the speed of light. In TET, gravity also occurs at the speed of light but appears to be instantaneous, as well. The reason for this is as follows: The transformations of t^+ to t^0, t^0 to t^-, t^- to t^0, and t^0 to t^+ each occur at the speed of light, which is the rate at which time becomes spacelike or space becomes timelike. The transformations of t^+ to t^0 and t^- to t^0 represent time becoming spacelike, with t^0 representing the spacelike state in both instances. The transformations of t^0 to t^- and t^0 to t^+ represent space, or really a spacelike state, becoming timelike. As described previously, gravity in TET involves the transformation of t^+ to t^0, t^0 to t^-, and the absorption of t^- by exterior space (see *Figure 3.42*); each of the previous three steps occurs at the speed of light. The absorption of t^- by exterior space also occurs at the speed of light, because it is essentially just the process of t^- transforming into t^0. However, because gravity cannot be said to have fully occurred until t^- is

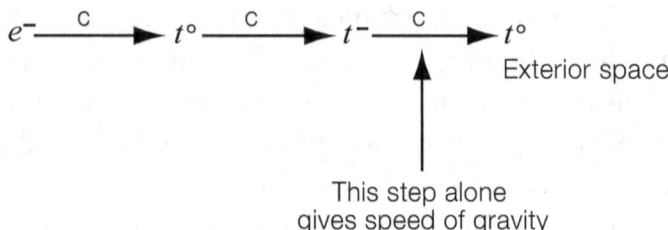

Figure 5.5 *Gravitational attraction can be regarded as occurring at the speed of light. Compare this figure with the electron schematic in Figure 3.42. c = speed of light.*

absorbed by exterior space, only this step actually plays a part in gravity's speed (*Figure* 5.5). Note that this does not mean that objects will necessarily move toward each other at the speed of light under the influence of gravity. Much of this motion is dependent on the strength of the bonds between temporal particles, which in turn is dependent on how tightly they are packed together.

The reason gravity appears to be instantaneous, even though it technically occurs at the speed of light is because the t^+ particles composing the gravitational field fill all of interior space. Consider two scenarios: In one of them, two particles interact electrically. In the other, two particles interact gravitationally. Again, let us make the particles magically appear in space. In the gravitational scenario, the two particles instantly interact with each other. This is because the gravitational field does not have to be built between them. The field exists on its own, independent of the two particles. In the other scenario, the particles cannot interact until an electric field has been built between them. The building of an electric field—the temporal polarization of interior space—also occurs at the speed of light. In the previous paragraph, it was stated that the speed of light is the rate at which time becomes spacelike and space becomes timelike. When green space transforms into blue and yellow space, it becomes timelike. Green space itself appears timeless. In transforming into blue and yellow lines, it takes on a definite timelike nature, with blue representing movement forward in time and yellow representing movement backward in time. Polarized lines of interior space, of course, do not preexist within that space; they must be built. Although the speed of light is fast, it is not instantaneous. Thus, whereas the particles in the first scenario

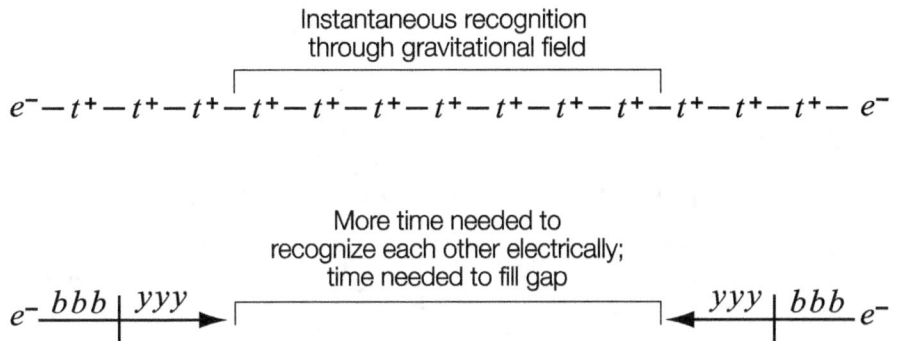

Figure 5.6 *Gravitational attraction can be regarded as occurring instantaneously. The gravitational field, unlike the electric field, preexists in space. It does not require time to be built between the electrons.*

instantly recognize each other's presence, the particles in the second scenario must wait until the field lines emanating from them meet to recognize each other's presence (*Figure 5.6*).

Spacetime Versus Space and Time

Whereas GR unites space and time into a single element called spacetime, TET disassociates them again. In TET, they work together but largely retain their individual identities. From the TET perspective, GR's view is simply an approximation of TET's. To help demonstrate this, consider a slice of curved GR spacetime connected to Earth. Consider this slice to be composed of a flexible tube filled with water, with the tube being space, water being time, and the entire system being spacetime. Now consider a small ball inside the tube at the top that slowly moves toward Earth. GR says that it is the curvature of the tube and water or the tube-water system (spacetime) that causes the ball to flow to the surface of the planet. TET, however, takes a step back; in this theory, the top of the tube is connected to a spigot, and Earth represents the drain. In TET, the ball flows down because the water flows down toward Earth. The spigot represents a white hole from which t^+ emerges, and Earth, the drain, represents a singularity into which t^- vanishes. This is an imperfect analogy because the shape and orientation of the curvature is not exactly in line with either theory, but it does well demonstrate an important difference

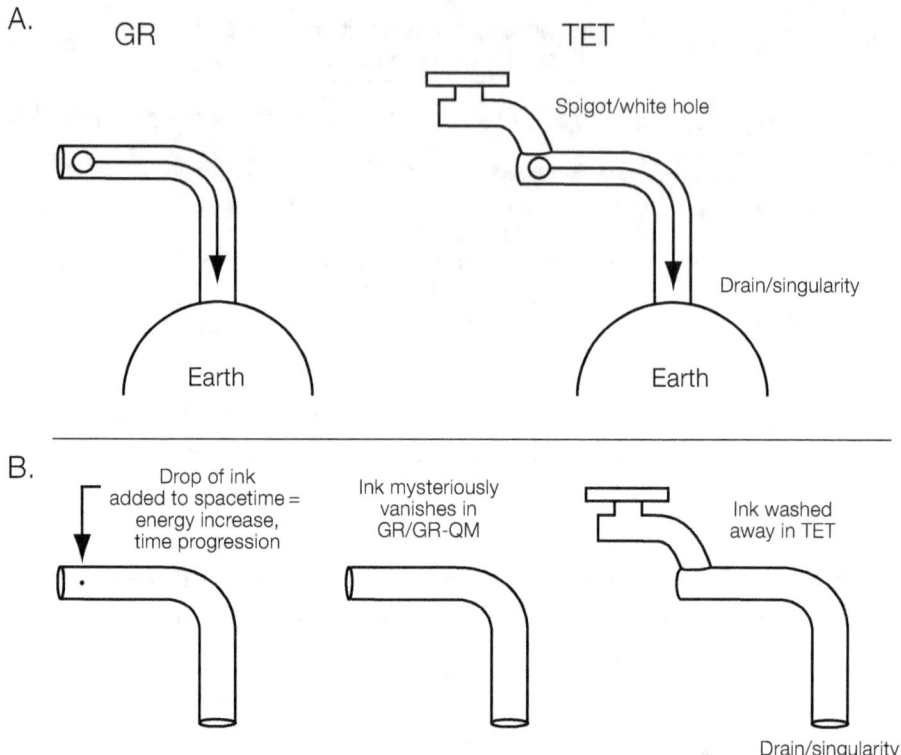

Figure 5.7 *(A) If a slice of curved spacetime in GR were likened to a curved water-filled tube, GR would say that the curvature of the tube-water system causes an object inside of it to move toward the surface of the Earth. TET suggests that it is the flow of the water (time) that causes the object to move in such a manner.*

(B) The quantization of GR would be like adding a drop of ink to the system. The ink represents a change (an increase) in the energy of the vacuum, and thus a change in time. In GR-QM, the ink mysteriously vanishes, and the system reverts back to its original energy state. Time never seems to move forward, leading to the problem of time. In TET, the energy is simply washed away in relation to the two opposing processes occurring within the tube—one adding water/time, and one removing it.

between GR and TET, which is that, in TET, it is not the curvature of space and time that causes gravity but rather deeper dynamics (*Figure 5.7A*).

This analogy also helps demonstrate why the quantization of GR—the combination of GR and QM into a single framework—leads to the problem of time, which again corresponds to the lack of energy changes in spacetime/the vacuum. Quantizing GR in this scenario is equivalent to adding a drop of ink to the tube-water system, representing a change in the energy level of the vacuum, which from the perspective of GR-QM represents a change in time. However, the ink mysteriously vanishes. In TET, it is simply washed away by the constant flow of water from the spigot to the drain (*Figure 5.7B*). Again, this is an imperfect analogy—this time because technically the ink should never be apparent in the system. It should be immediately washed away upon being added.

Curvature of Light in a Gravitational Field

TET is not needed to understand the curvature of light around a massive body. Light only knows straight lines. Thus, as space is curved around the body, light too will bend as it passes in the vicinity of that body. From the TET perspective, the temporally polarized lines composing the light simply form in that curved space as the light propagates, continuing on a straight path thereafter, if no other massive body is present.

CHAPTER
6

QUANTUM MECHANICS REVISITED
AND SPECIAL RELATIVITY

This chapter revisits and introduces new Quantum Mechanics (QM)–related ideas, as well as introduces ideas related to Special Relativity (SR), discussing them from the TET perspective. In such a discussion, it is helpful to begin with the concept of vacuum. While a vacuum in QM is just space, a vacuum in TET is technically space and time. SR's vacuum is also space and time, but as in General relativity (GR), they are united into spacetime. Whereas spacetime is curved in GR, it is flat in SR. Like QM, SR appears to view the universe from TET's t^-/t^+ perspective. This is because, generally, curved space represents space that is absorbing temporal energy (by way of t^+/t^-), whereas flat space represents space that is emitting temporal energy (by way of t^-/t^+). QM and SR often are united into a single framework, here called QM-SR. From the perspective of TET, this is doable because they each view the universe from the same temporal perspective (t^-/t^+). The principal difference between QM and SR from the standpoint of TET is that SR focuses solely on the t^+ emitted rather than the whole t^-/t^+ process. SR thus starts from the standpoint of flat space filled with t^+.

General Relativity and Special Relativity

As GR deals with mass, SR deals with motion, producing similar changes to a moving object's size, time, and mass as GR does to a massive object's size, time, and mass. Recall the idea that temporal particles are attracted to all forms of energy. Thus, when an object moves through space, temporal particles move toward it due to this motion energy. Although they surround the object, they principally move in from the front, in direct line of the object's motion, meeting it head on—note the temporal particles offer no resistance to the movement. As they move toward the object, they compress the space surrounding it and the space that is the object (that is, its frame, or frames if a composite), decreasing its size, slowing its time, and increasing its mass, as described in the previous chapter regarding compression by a gravitational field. The faster the object moves, the more temporal particles will surround and compress it, altering its size, time, and mass.

Because temporal particles move in mostly from the front, the object will appear to shrink more from that direction. An object moving at a constant velocity will maintain the corresponding level of compression until it either slows down or moves faster. As it slows down, some temporal particles will dissipate, and the object's size will become larger. Its time will increase, and its mass will decrease. The object itself and other objects moving in unison with it will notice nothing out of the ordinary about their size, time, and mass at any point while moving at a constant rate, slowing down, or moving faster together. The size, time, and mass changes would only be measurable from someone outside of their compressed area, which is the same as saying outside of the density of temporal particles they are immersed in—the temporal particles that moved in because of their motion—which in turn is the same thing as saying outside of the objects' reference frame.

Thus as noted in the previous chapter, the reference frame of an object, from the perspective of TET, is the density of temporal particles it is immersed in. It may be the temporal particles surrounding it due to its mass or motion. It could also be the temporal particles surrounding it due to another object's mass (such as that of Earth) or motion (such as that of an airplane). Because the density of temporal particles that objects find themselves in can vary dramatically because of differences in mass and motion, there is no "correct" value for size, time, or mass. These have only relative values.

SR plays an important role in QM, in that SR makes virtual force carriers real. From the TET perspective, and as explained in chapter 3, this occurs by compression of field lines—electric, weak, or strong—when they form a wave pattern. As waves, they represent typical photons (electric force), Z bosons (weak force), or gluons (strong force). When in a compressed state, they are able to stand out from the vacuum. With enough energy, photons can give rise to, for instance, electrons and positrons, Z bosons to electron neutrinos and anti–electron neutrinos, and gluons to up quarks and anti–up quarks. Currently, SR is considered to play a role in the creation of virtual particles as well. However, in TET, virtual particles are simply static field lines and their associated temporal particles. Thus, the compression associated with SR is not needed to create virtual particles in TET.

Magnetism and Special Relativity

Magnetism is largely an SR phenomenon. As objects move through space, temporal particles converge on them. If they are moving parallel, they will undergo magnetic attraction; if they are moving anti-parallel, they will undergo magnetic repulsion, as described previously. In TET, an important reason magnetic attraction under relativistic conditions is able to overcome electric repulsion is that as two electrons, for example, move through space and temporal particles move in and compress them, the number of field lines emanating from them is reduced. As described in the previous chapter, when the frames of electrically charged particles are compressed, they have fewer temporal spaces, which corresponds to the frames having fewer field lines emanating from them. With fewer field lines, the repulsion between the electrons is reduced (*Figure 6.1*).

Heisenberg Uncertainty Principles

The Heisenberg uncertainty principles are a cornerstone of QM. One of the principles states that it is impossible to know both the position and momentum of an object with a high degree of certainty. The more you know about an object's position, the less you know about its momentum and vice versa. From TET's perspective, the reason this occurs is because to gain momentum information, you literally have to move away from position information and vice versa. For example, consider an electron moving through space. The electron is composed of a frame, mass field, and spin field. Because it is moving, there is an extra volume of temporal energy surrounding the electron due to its motion energy. This volume of temporal energy and the volume that is its

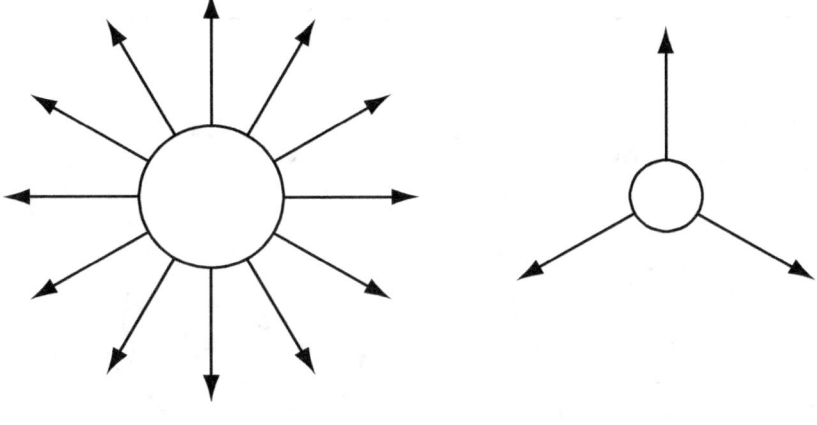

Less compressed electron   More compressed electron

Figure 6.1 *When the frames of electrically charged particles are compressed, fewer field lines emanate from them. If two electrons are moving through space, the fewer field lines lead to a reduction in electric repulsion, aiding their magnetic attraction if they are moving in a parallel manner (see Figure 3.18). If an electron is in a metal, for example, and becomes compressed following excitation, the reduction in its field lines may allow it to breakaway from the metal. In this case, the reduction in field lines leads to a weakening of the attractive forces holding onto the electron.*

mass field together constitute its momentum field or simply its momentum. As shown in *Figure 6.2*, to gain the most precise information you can about the particle's position, you have to move toward its frame, which is away from momentum information. To gain the most precise information you can about the particle's momentum, you have to move away from its position information. Because you cannot move toward the particle frame and away from it at the same time, you cannot know both its position and momentum with high precision.

This uncertainty even applies to a stationary particle. Such a particle does not have an extra volume of temporal particles surrounding it because it is not moving, not including the gravitational field energy. However, its mass is part of its momentum. To gain the most information you can about mass, as with momentum in general, you have to move away from the particle's frame and thus lose information about its position. The more you move toward the particle's frame, the more you know about its position, but the less you know

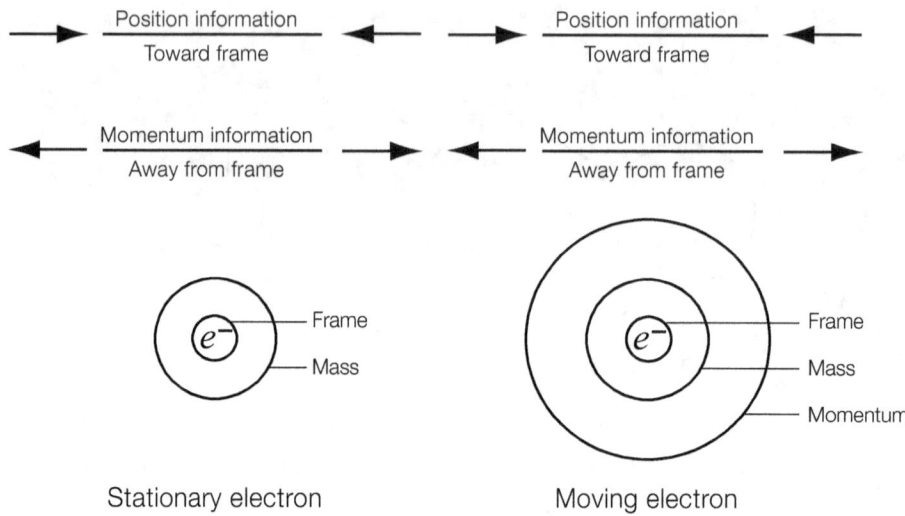

Figure 6.2 *To gain the most precise information you can about a particle's position, you have to move toward its frame, which is away from momentum information. To gain the most precise information you can about a particle's momentum, you have to move away from its position information. Because you cannot move toward the particle frame and away from it at the same time, you cannot know both its position and momentum with high precision.*

about its mass and thus its momentum (*Figure 6.2*). Note that because our experience is with mass only, in that we never deal directly with a particle's frame because the frame is shielded by the particle's mass, we can never be sure where exactly any QM particle or composite structure is at any moment in time—this applies to everything from an individual electron to a book sitting on a person's desk, to planets, stars, galaxies, and beyond. We can only speak of the probability of finding the object in a particular location in space. It may be a high probability, but a probability nonetheless. Because of an object's mass, we are prevented from ever knowing where exactly it is.

The other uncertainty principle involves time and energy. According to the principle, you cannot know both the time and energy of an object with a high degree of precision. The more you know about an object's time, the less you know about its energy. The more you know about its energy, the less you know about its time. In TET, the reason for this is the same reason momentum and position information cannot both be known with high precision.

That is, in order to get time information, you have to move away from energy information and vice versa. Using an electron as an example again, this is because the greatest time information resides on the particle's frame (where time is processed), whereas the greatest energy information is gained by knowing its mass and motion, which requires you to move away from the particle frame. Because you cannot move toward the particle frame and away from it at the same time, you cannot know both its time and its energy with high precision.

Speed of Light as Limiting Speed

From the perspective of TET, the speed of light is the limiting speed in the universe because the frames of all particles originate from light or its derivatives (Z bosons/gluons). For a particle such as an electron to move faster than light, it would have to move faster than the ability of interior space to generate an electric field line, which in the case of the electron is its frame. The electron would thus be moving faster than it can exist in interior space. It would be moving through interior space faster than the ability of interior space to generate it, which is impossible.

CHAPTER

7

ATOMIC THEORY

All of the materials of our everyday existence are built from electrons, up quarks, and down quarks (generation I matter, see *Tables* 3.5 and 3.8). Basically, up quarks and down quarks combine to form protons and neutrons, which join with each other to form atomic nuclei, which in turn join with electrons to form full-fledged atoms. Elements such as hydrogen, oxygen, lead, gold, and silver are, at their fundamental level, different types of atoms. One of the major differences between TET and contemporary theories of the atom is that in TET, electrons are considered to be in fairly fixed positions around the nucleus of an atom. They are not considered to orbit the nucleus. The natural question then is, Why do the electrons, being negatively charged, not crash into the nucleus, which is positively charged? Why do they stay so far from the nucleus despite the electric force, which causes them to be attracted to the nucleus?

Recall several previous ideas:

1. The initial polarized lines emanating from a particle such as an electron mimic the particle; that is, they become a reflection of the particle itself, just in the form of a line.

2. When particles bond with each other, they act as a single particle. More precisely, their frames act as one frame.

3. Any particle with three excess yellow segments, such as the proton, will have one unit of positive charge. What is meant by excess is that there are no segments of opposite polarization in the particle frame to essentially cancel out the segments' effects.

4. Any excess weak charge within composite particles like protons are in a sense buried within them. All that is visible from the outside is the electric charge, or lack thereof, of their frames.

5. The field lines emanating from electrically charged composite particles are the same as those generated by electrons and positrons. In the case of protons and similar particles, however, this is only the case some distance from those particles. The areas of the field lines closest to the particles are a little more complicated.

The frame of a proton is composed of four yellow segments and one blue segment. The frame of a composite particle, unlike the frame of an elementary particle or individual primary or secondary field line, can have polarized segments that are starkly opposite (e.g., yellow and blue). Because initial field lines mimic the frame of their source particle, the proton would normally be surrounded by lines with four yellow segments and one blue segment; however, because individual primary or secondary field lines cannot have starkly opposite segments, each of the proton's initial lines fluctuates between one state having four yellow segments plus one bluish-green segment and another state having four yellowish-green segments plus one blue segment, with its sister line doing the same but oppositely (*Figure 7.1A*), and with each primary and secondary line thereafter doing the same, *up to a point*. That is, the radiating lines have these natures only close to the proton's frame. Further away, the field lines stabilize into the typical format for a positively charged particle (three yellow segments in the primary line followed by three blue segments in the secondary line). This is because, irrespective of the fluctuations of the field lines, there is an extra blue and yellow segment being generated in the same direction within a single line. Close to the proton's frame, they remain (and of course fluctuate between electric and weak states). However, once some distance has been gained from the proton's frame, there is no incentive for interior space to give the extra blue segment and yellow segment independent existence. They fade to green, with the remaining field line containing only three yellow segments and its sister three blue segments (*Figure 7.1B*).

When an electron bonds with a proton, electric attraction occurs normally up to the point where the proton's field lines become complicated. At this point,

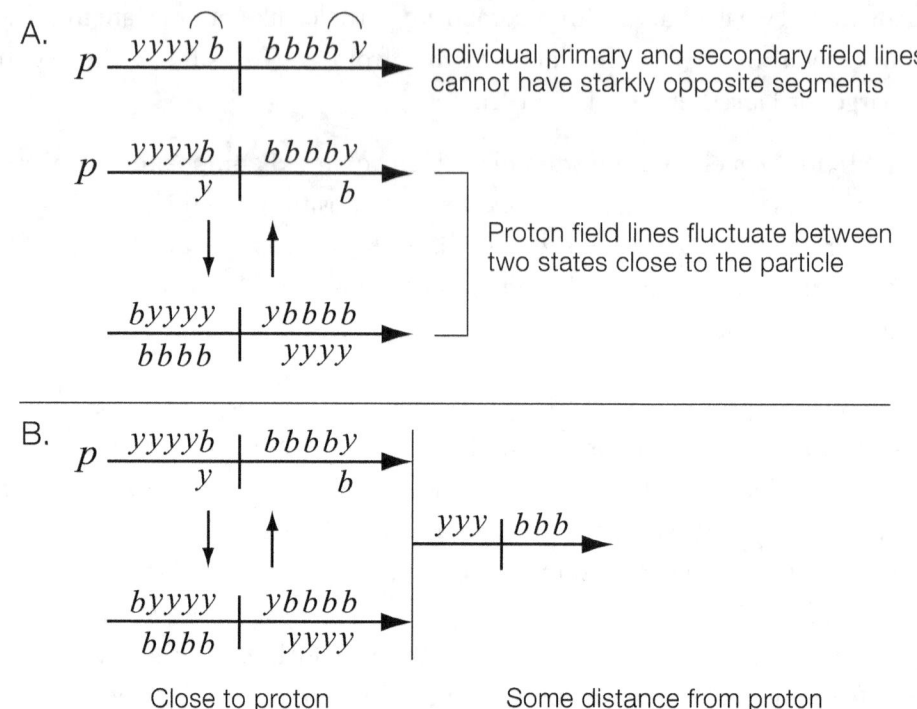

Figure 7.1 *(A) The initial lines emanating from protons fluctuate between one state having four yellow segments plus one bluish-green segment and another state having four yellowish-green segments plus one blue segment, with the sister lines doing the same but oppositely. This continues down the line for some distance beyond the proton's frame.*

(B) The field lines stabilize into the typical format for a positively charged particle (three yellow segments in the primary line followed by three blue segments in the secondary line).

the electron cannot move closer to the proton because it is electrically blocked. Ultimately, it is blocked because the proton has a slightly negative nature, owing to the blue segment in its frame, which is translated for some distance down the field lines emanating from it. Also, with the field lines close to the proton fluctuating between two different states, the electron position shifts slightly. One state allows the electron to get closer to the proton, while the other pushes it back (*Figure 7.2*). The result is that the electron remains some distance from the proton. It does not fall away from the proton, because there is still electric attraction between it and the proton, but it

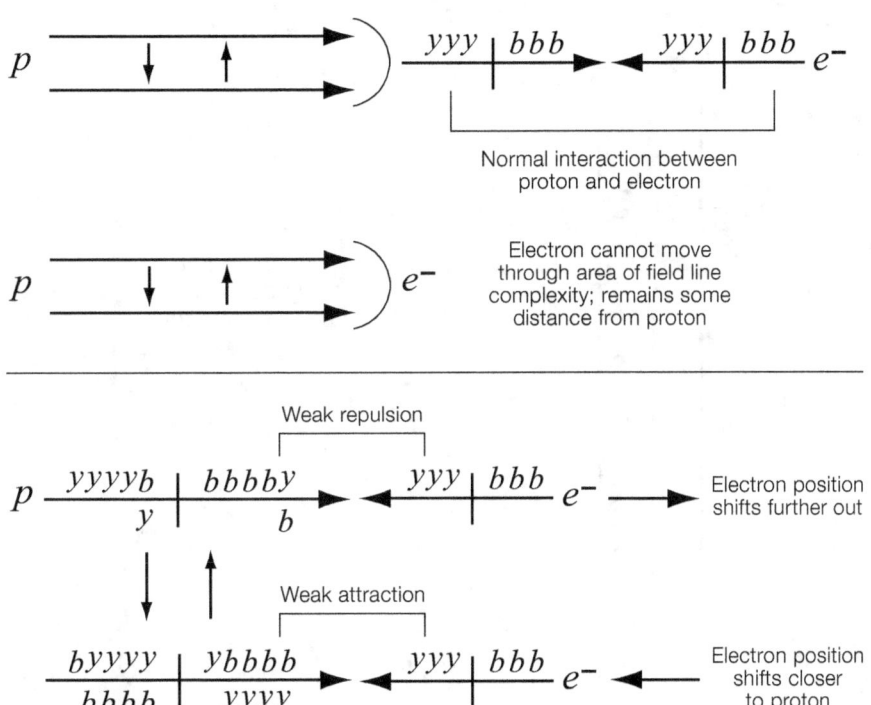

Figure 7.2 *An electron and proton experience typical electric attraction up to the point where the proton's field lines become complex. The proton has a slightly negative nature, which manifests itself closer to the particle by way of its complex field lines. This negative nature of the proton repels the electron, preventing the electron from falling fully into the proton. With the proton's complex lines fluctuating between two states, the electron's position shifts slightly from time to time.*

cannot move through the area of complexity further toward the proton because its field lines and those of the proton will not cancel fully.

A similar scenario exists for the neutron; that is, it has complicated field lines close to its frame. However, because the neutron is ultimately generating two yellow and two blue segments in the same direction within an individual field line, interior space reduces them to true green some distance from the composite particle. Because all of its field lines eventually disappear (that is, turn true green) some distance from the neutron, it appears electrically neutral.

The more protons there are within a nucleus, the more electrons there will be surrounding it. The electrons arrange themselves in the most electrically and

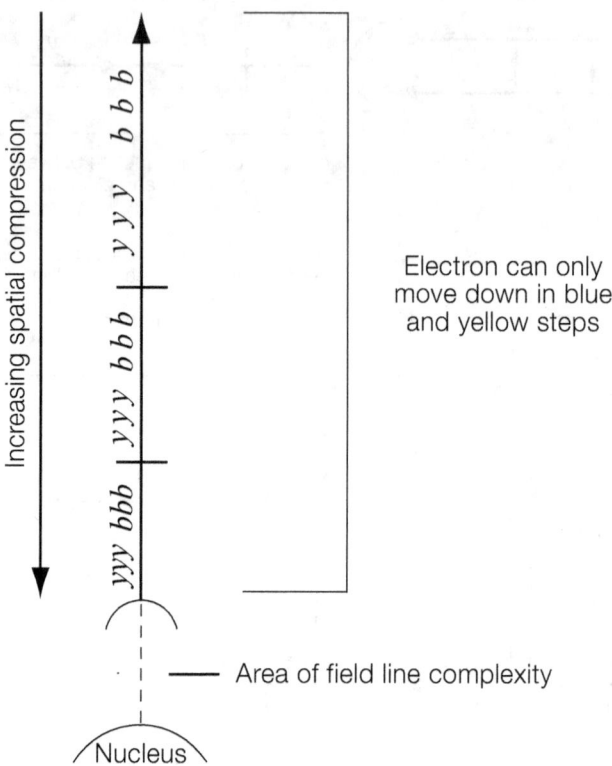

Figure 7.3 *When electrons move toward or away from a nucleus, they can do so in blue and yellow steps only, corresponding to discrete energy levels around the nucleus.*

magnetically favorable positions, giving rise to such regions around the nucleus as energy levels (or energy shells) and orbitals. Energy levels relate to the electrons' distance from the nucleus and thus their potential energy. Changes in the potential energy of electrons can occur only in discrete amounts; that is, changes in the distance of the electrons from the nucleus do not occur smoothly. In TET, the discrete, as opposed to continuous, changes in distance correspond to the blue and yellow natures of the electric field lines connecting the nucleus and the particular electron. When two particles move closer to each other along an electric field line or even if they move further away from each other, they essentially do so in blue and yellow field line "steps," spending little time between these steps, although "substeps" may arise due to the individual blue and yellow segments within the blue and yellow lines

(*Figure* 7.3). The steps and substeps are discrete energy levels, with the main idea being that electrons can exist only at particular levels of potential energy around a nucleus. For an electron to move to a higher energy level, further from the nucleus, it must absorb an amount of energy equal to the difference between the two levels. Similarly, when an electron moves to a lower energy level, closer to the nucleus, it must release an amount of energy equal to the difference between the two levels.

Orbitals are areas around the nucleus where a particular electron spends most of its time. A single orbital can hold a maximum of two electrons; the

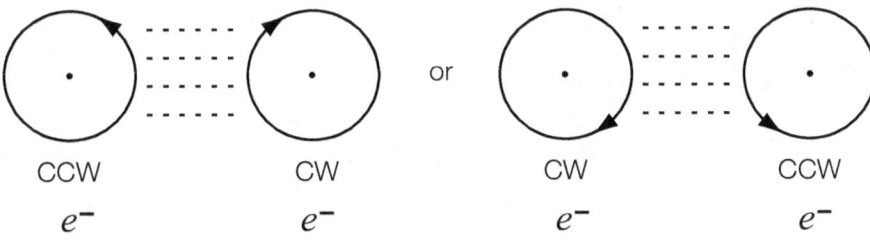

CCW   CW   or   CW   CCW
e^- e^- e^- e^-

Figure 7.4 *In TET, an orbital is simply any area around a nucleus where two electrons with opposite spins can exist comfortably.*

electrons must, however, have opposite spins to remain close to each other. The opposite spins allow the electrons to attract each other magnetically, which helps stabilize them under the stress of electric repulsion (*Figure* 7.4). Earlier, it was stated that the source of the magnetism of bar magnets comes from the spin of electrons; such electrons must be unpaired in their respective orbitals. Basically, macroscopic magnetism occurs when the unpaired electrons are arranged in such a way that their individual spins add up (*Figure* 7.5).

Note that protons and neutrons (called nucleons generally) can also have a magnetic nature. From the TET perspective, the overall spin of a proton or neutron will be the net spin within that composite particle (*Figure* 7.6). Although three electrons cannot occupy a single orbital, three quarks can exist within a

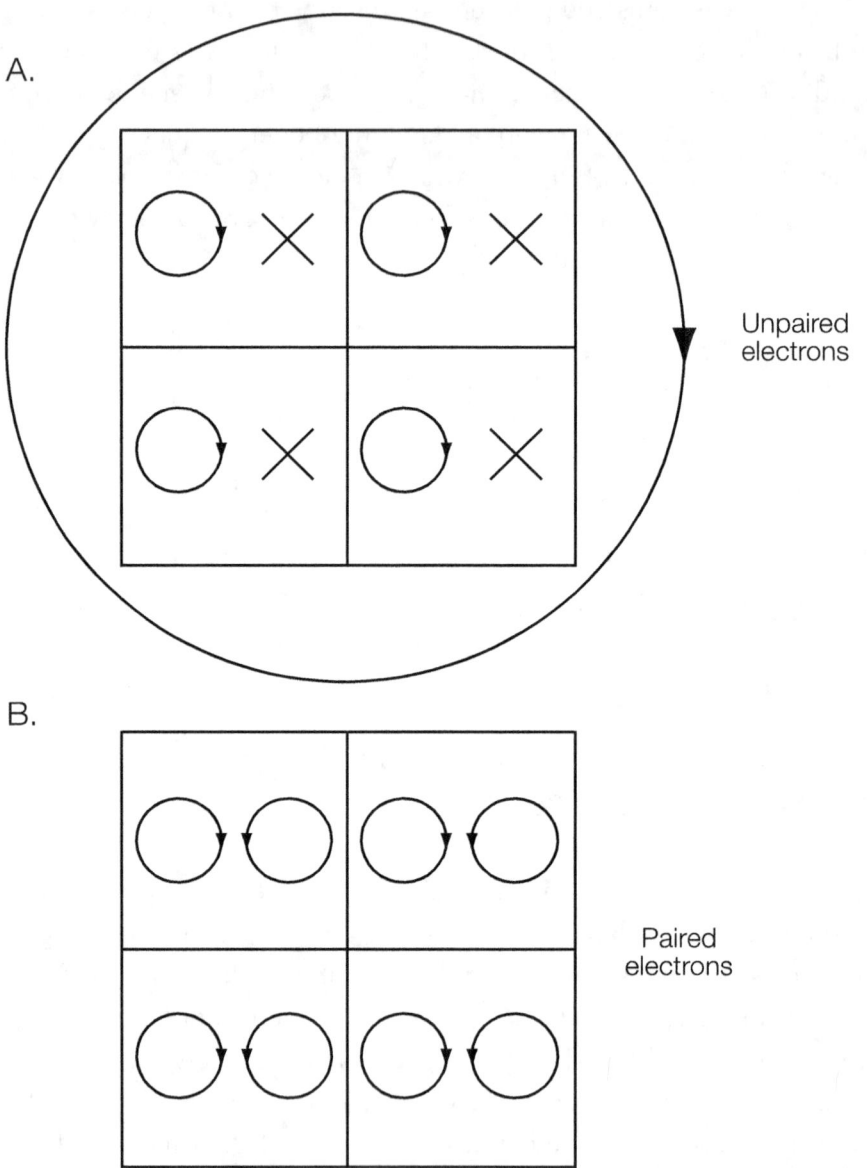

Figure 7.5 *(A) For macroscopic magnetism to occur, electrons must be unpaired in their orbitals and arranged in such a way that their individual magnetic natures add up to produce a larger effect.*

(B) When electrons are paired, the magnetic nature of one has the effect of canceling the magnetic nature of the other, such that no macroscopic magnetism occurs.

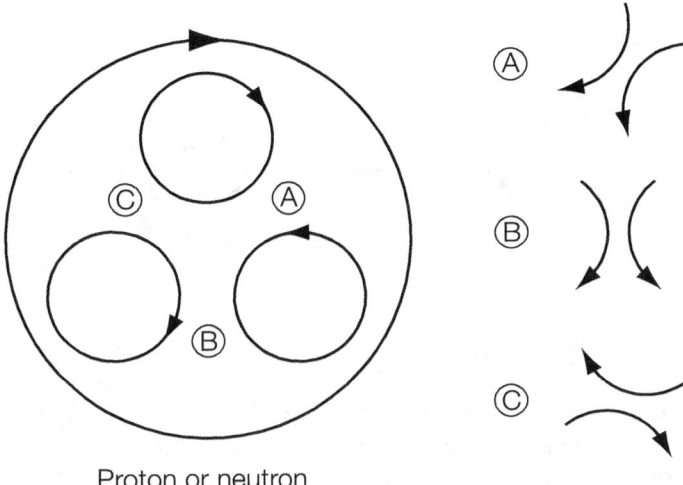

Proton or neutron

Figure 7.6 *Protons and neutrons can also have a magnetic nature. From the TET perspective, the overall spin of a proton or neutron will be the net spin within that composite particle. In the figure, two quarks have clockwise spins and one has a counterclockwise spin. The overall particle thus has a clockwise spin. Note that there is more magnetic attraction between the quarks (at areas A and B) than there is magnetic repulsion (at area C).*

proton or neutron because while the opposite spins drive some of the quarks apart, the strong force keeps the composite particle together and the nuclear contact force, a physically strong force at short range (discussed below), holds the entire nucleus (protons plus neutrons) together.

The nuclear contact force is said to be residual effects from the strong force holding quarks together. In TET, this force is born from a symbiotic relationship between a blue segment on one nucleon and a yellow segment on another. Basically, a blue segment bonds with a t^+ particle, transforms it to t^-, and then repels the particle. A yellow segment, working with this blue segment, bonds with the repelled t^- particle, transforms it to t^+, and then repels it, at which point the blue segment bonds with it again. Of course, some t^- is always being lost to exterior space, but essentially, a blue segment and a yellow segment

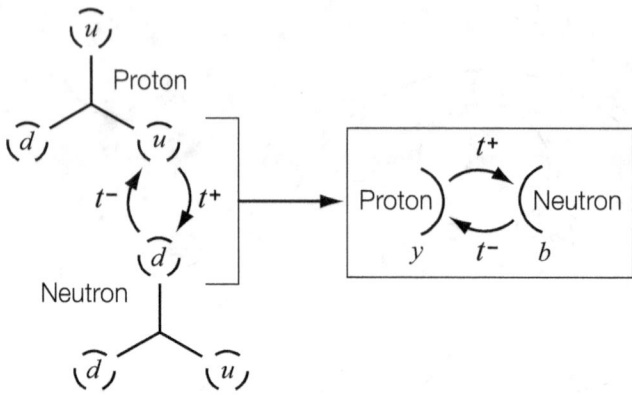

Figure 7.7 *In TET, the nuclear contact force is born from a temporally symbiotic relationship between a blue segment on one nucleon and a yellow segment on another, in which the segments exchange t$^+$ and t$^-$ particles. The interacting nucleons can also be two protons or two neutrons.*

bounce t$^+$ and t$^-$ particles back and forth, pulling and pushing those particles, and in so doing hold the nucleons they are a part of together. For this to occur, the nucleons have to be very close to each other, nearly in contact (*Figure 7.7*). (The force's name is a bit of a misnomer, as the nucleons are not necessarily touching.) The strength of the force is enough to override the repulsive force between protons, allowing many protons and neutrons to remain close to each other. When the nuclear contact force is broken, the protons fly apart, perhaps taking some neutrons with them.

CHAPTER
8

PARTICLE PHYSICS

A ll of physics is essentially particle physics, but this chapter focuses on the transformation of matter and anti-matter particles into different matter and anti-matter particles, sometimes referred to as particle decay. As noted earlier, the transformation may occur through the process of repolarization, in which the frames of particles are blue shifted or yellow shifted. In order for the segments within the frame of a particle to shift their polarization, the particle must partner with another particle whose frame segments will be shifted oppositely. Particles with at least one weak (bluish-green or yellowish-green) segment within their frames require less energy to shift their polarization than a particle with a solely electric frame. Therefore, particles often partner with neutrinos or quarks to shift their polarization.

Neutrinos, being bluish-green, can only be blue shifted, whereas anti-neutrinos, being yellowish-green, can only be yellow shifted. Up, charm, and top quarks, having fully yellow segments, can only be blue shifted. Down, strange, and bottom quarks, having fully blue segments, can only be yellow shifted. The nature of the principal particle being transformed, typically an unstable particle, will dictate which type of neutrino or quark it will interact with.

Figure 8.1 illustrates the decay of a muon and anti-muon. The neutrino/anti-neutrino pairs in this instance are manufactured through the energy of the two systems—energy present due to the instability of the muon and anti-muon. However, a muon could react with a stray electron neutrino that just happened by, in which case the anti–electron neutrino would not be present in the reaction. Likewise, the anti-muon could react with a stray anti–electron

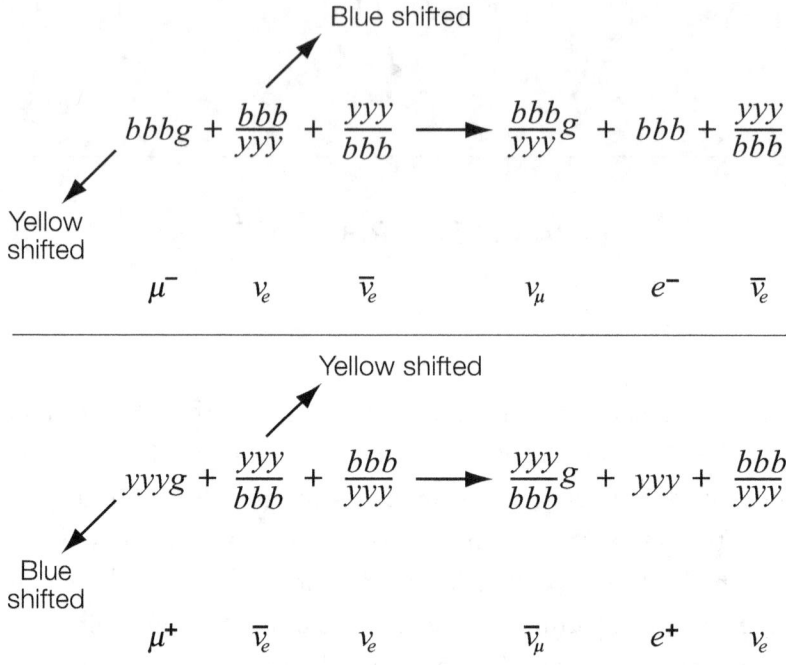

Figure 8.1 *In TET, a muon can decay be inducing the production of an electron neu-trino/anti–electron neutrino pair and interacting with the electron neutrino, such that the muon is yellow shifted, becoming a muon neutrino, and the electron neutrino is blue shifted, becoming an electron, with the opposite occurring for an anti-muon.*

neutrino that just happened by, in which case the electron neutrino would not be present in the reaction (*Figure* 8.2).

Quarks may also decay via repolarization. For example, when combined with one or more protons in a nucleus, neutrons are usually very stable. The protons, in a sense, help settle them down. If there are too many neutrons in a nucleus, however, when the balance between neutrons and protons is upset, the insta-bility causes the nucleus to jostle about. The field lines present in the nucleus will fluctuate wildly as a result, producing at times neutrino/anti-neutrino pairs. With enough energy, an electron neutrino can interact with one of the down quarks in a neutron, transforming the down quark into an up quark—and thus the neutron into a proton—while transforming itself into an electron. The down quark would be yellow shifted and the electron neutrino would be blue shifted during the process, bringing stability to the nucleus (*Figure* 8.3).

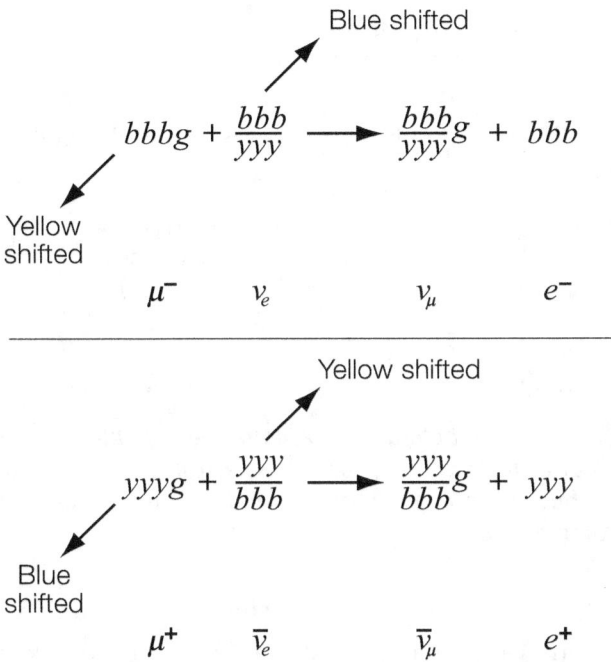

Figure 8.2 *Instead of inducing the creation of a neutrino (and an anti-neutrino), a muon may decay by interacting with a stray neutrino already in existence. An anti-muon may decay by interacting with a stray anti-neutrino.*

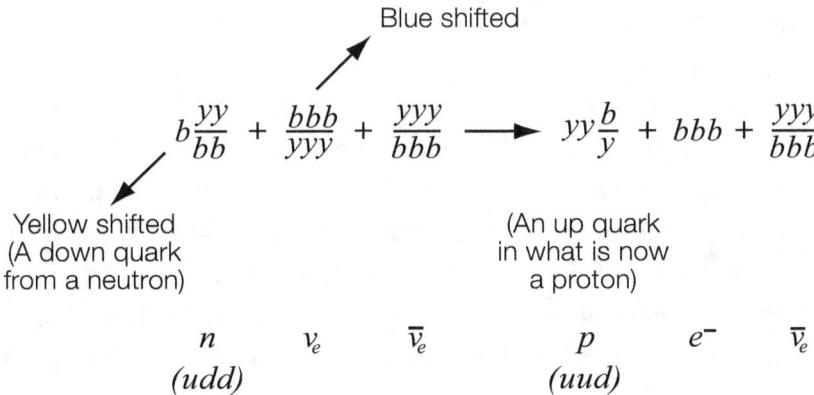

Figure 8.3 *In one form of beta-decay, a neutron decays into a proton. In TET, an electron neutrino/anti–electron neutrino pair is created, and one of the down quarks of the neutron interacts with the electron neutrino, with the down quark being yellow shifted, becoming an up quark, and the electron neutrino being blue shifted becoming an electron. With the down quark changing into an up quark, the nucleon transforms from a neutron into a proton.*

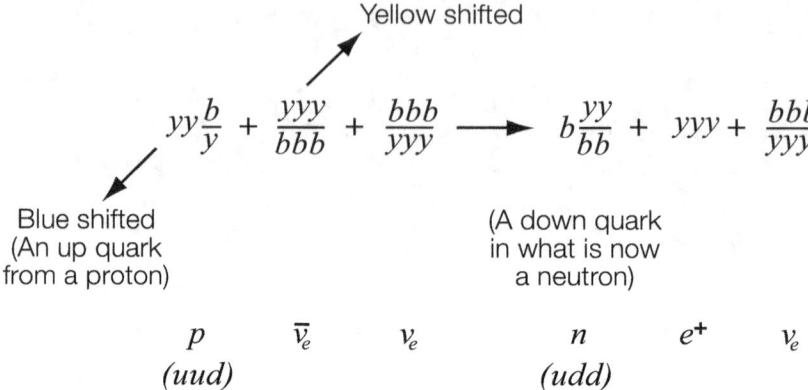

Figure 8.4 *In the other form of beta-decay, a proton decays into a neutron. One of the up quarks in the proton interacts with an anti–electron neutrino, with the up quark being blue shifted, becoming a down quark, and the anti–electron neutrino being yellow shifted becoming a positron.*

If a nucleus has too many protons, the reverse may occur. In this case, the anti–electron neutrino would interact with one of the up quarks in the proton, transforming the up quark into a down quark—and thus the proton into a neutron—while transforming itself into a positron (*Figure 8.4*). This transformation and the previous one are called beta-decays and could also occur by a stray electron neutrino interacting with a neutron or a stray anti–electron neutrino interacting with a proton.

With enough energy within a system, neutrinos of higher mass may be created and aid in repolarization. *Figure 8.5* shows the decay of a negative pion into a neutral pion, muon, and anti–muon neutrino and the decay of a positive pion into a neutral pion, anti-muon, and muon neutrino. Note that the decay modes most frequently observed do not include neutral pions. Being very unstable, it is likely that each neutral pion decays quickly into photons (meaning neutral photons, but actually gluons in TET), adding their energy to the kinetic energy of the remaining particles.

In TET, there is a process called "green segment aggregation," in which particles with green segments can jettison those segments as long as a sterile neutrino can be formed (*Figure 8.6*). Shown is the decay of a lambda particle, composed of an up quark, a down quark, and a strange quark. The lambda

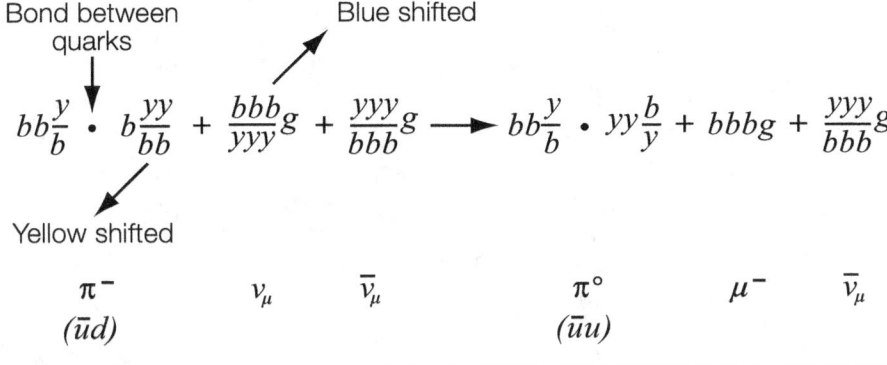

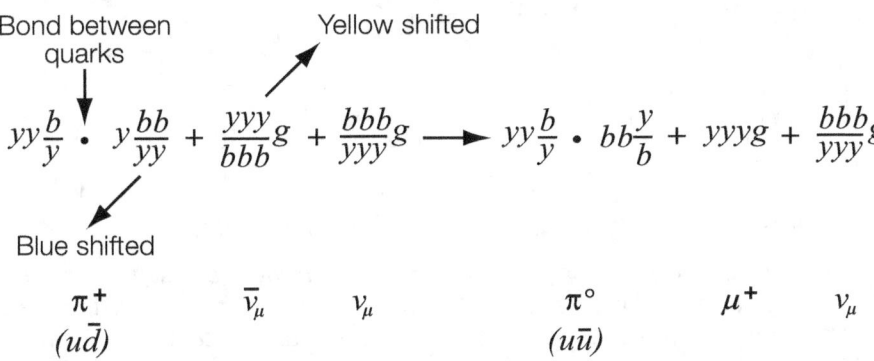

Figure 8.5 *With enough energy within a system, neutrinos of higher mass may be created and aid in repolarization. In the examples shown, a negative pion and a positive pion induce the creation of a muon neutrino/anti–muon neutrino pair. Note that either quark within the pions may be repolarized. If the anti–up quark within the negative pion and the up quark within the positive pion were repolarized, the neutral pions produced in each example would be composed of a down quark and an anti–down quark, instead of an up quark and an anti–up quark. Note also, that the final products of negative pion decay could be just the muon and anti–muon neutrino and the final products of positive pion decay could be just the anti-muon and muon neutrino if the neutral pion in each case quickly decayed into a photon (meaning a neutral photon, but actually a gluon in TET) and added its energy to the kinetic energy of the other particles.*

particle is unstable and through the energy of this instability can induce the creation of a pair of particles with green segments in their frames—particles high in mass. But in this case, instead of the pair being a neutrino and an anti-neutrino, it is a quark and an anti-quark (specifically, a strange quark and an anti–strange quark). Note that the two quarks are linked; no isolated quarks have ever been found in nature.

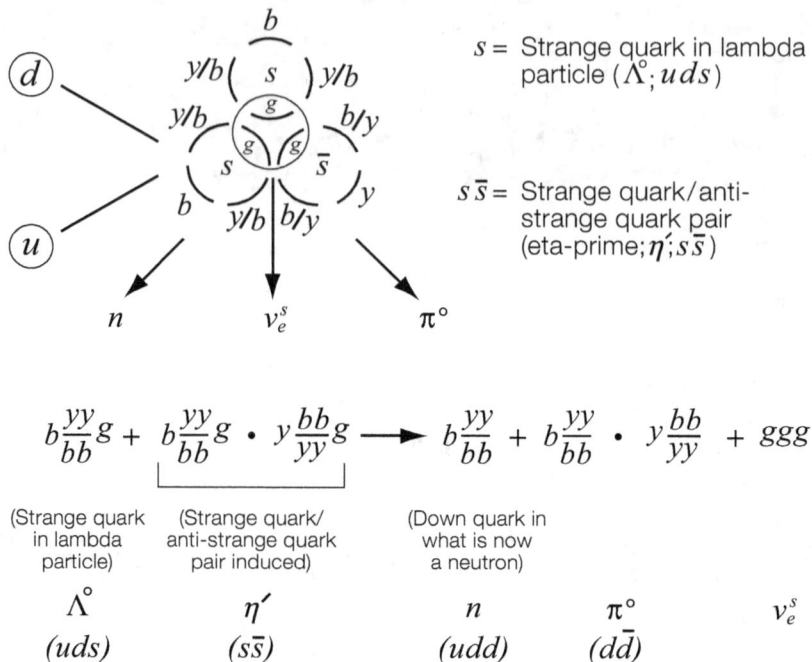

$$b\frac{yy}{bb}g + b\frac{yy}{bb}g \cdot y\frac{bb}{yy}g \longrightarrow b\frac{yy}{bb} + b\frac{yy}{bb} \cdot y\frac{bb}{yy} + ggg$$

(Strange quark in lambda particle)	(Strange quark/ anti-strange quark pair induced)	(Down quark in what is now a neutron)		
Λ°	η'	n	π°	v_e^s
(uds)	*(s$\bar{s}$)*	*(udd)*	*(d$\bar{d}$)*	

Figure 8.6 *In green segment aggregation, particles with the green segments can jetti-son those segments as long as a sterile neutrino can be formed. In the example shown, the lambda particle loses its green segment and becomes a neutron. The eta-prime particle loses its green segments and becomes a neutral pion.*

As shown in the figure, the lambda particle interacts with the strange quark/anti–strange quark pair (eta-prime). Through the interaction, the green segments from the two strange quarks and the anti-strange quark combine to form one sterile neutrino. Through this action, the lambda particle transforms into a neutron, and the strange quark/anti–strange quark pair transforms into a neutral pion. TET's sterile neutrinos have not been experimentally proven; indeed, if a sterile neutrino were present, it would be difficult to know as they are very hard to detect, having no electric or permanent weak charge. Also, they have a random spin and would appear massless, or very close to it. They would largely blend in with the vacuum. They should possess momentum, however, and be able to alter the position and momentum of another particle they bump into.

Repolarization and green-segment aggregation may occur together. For example, not only may the lambda particle get rid of the green segment in its strange

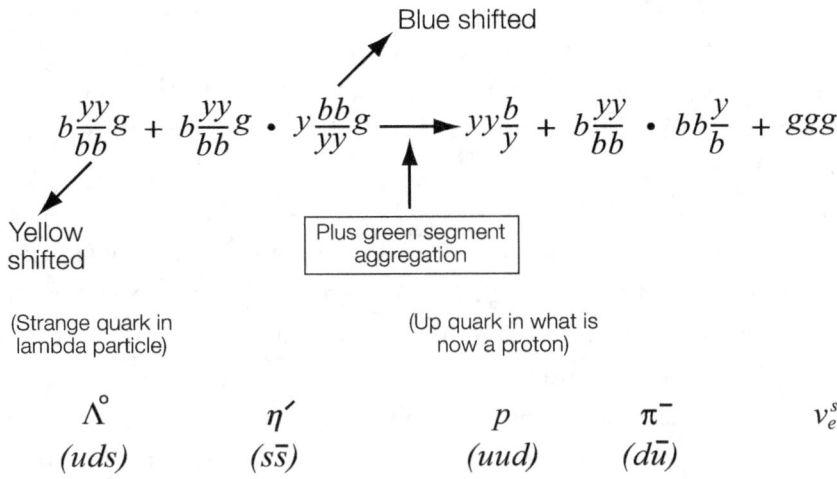

Figure 8.7 *Repolarization and green segment aggregation may occur together.*

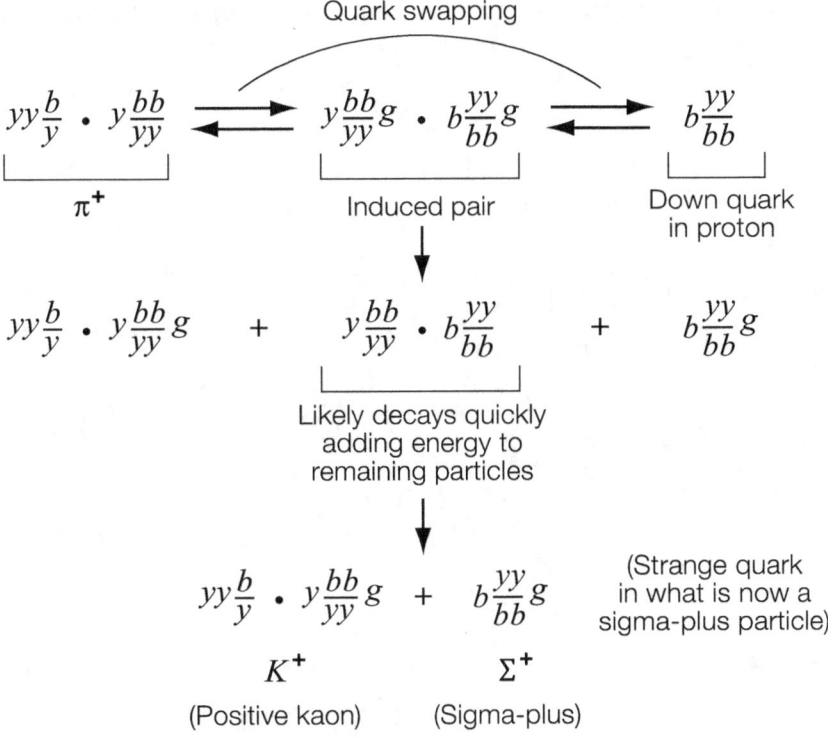

Figure 8.8 *Quark swapping is another decay mode in TET. In this reaction, a composite structure can only release a quark if it can latch onto another quark simultaneously.*

quark, but that quark may also undergo repolarization due to its instability (*Figure* 8.7). With repolarization and the ejection of its green segment, the strange quark transforms into an up quark, and thus the lambda particle transforms into a proton. Also created in this decay is a negative pion and a sterile electron neutrino.

Another TET decay process is quark swapping, which like repolarization may or may not occur with green-segment aggregation. *Figure* 8.8 shows this phenomenon occurring in the interaction between a positive pion and a proton. A high-energy pion can induce the creation of a strange quark/anti–strange quark pair. The pion exchanges its anti–down quark for the anti–strange quark of the pair, and the proton exchanges its down quark for the strange quark in the pair. In the end, the positive pion is transformed into a positive kaon, and the proton is transformed into a positive sigma particle. The newly formed neutral pion decays into what would be identified as photons, but are gluons from the perspective of TET, adding their energy to the energy of the remaining particles.

CHAPTER
9

COSMOLOGY

This chapter examines TET's stance on the universe as a whole, including development. From the standpoint of TET, the universe began as a point-like, volumeless, primordial seed of green space, in a frozen-like state, teetering on the razor-thin edge of its blue and yellow natures. The space is then considered to have undergone spontaneous* temporal polarization, in a sense thawing out, developing basically into two spaces: one blue and one yellow. The polarization also likely jostled the spaces, causing them to expand slightly. This polarization and initial expansion together represent the "Big Bang" in TET. With the expansion, the spaces developed volume but contained nothing. That is, they were true vacuums—devoid even of time. Developing into a true vacuum excites a space; it energizes it. Temporal particles are created within that space to bring the energy level down. With temporal particles likened to water, this is analogous to an expanded balloon that is filled with water being more stable than one that is not on Earth. The waterless balloon pops very easily compared with the water-filled balloon. That is, the waterless balloon— i.e., space devoid of temporal particles—is in a more excited state. Note that in interior space, which is blue at this point, t^+ particles are created, whereas in exterior space, which of course is yellow, t^- particles are created.

The creation of temporal particles in a true vacuum relates to the General Relativity idea that the level of energy in the universe remains constant over time. The expansion of the two spaces creates true vacuums, which raises the energy of the universe, whereas the creation of temporal particles within those spaces lowers the energy. Up until this point, these two processes have been described as occurring in tandem, one then the other, but should actually be considered to occur simultaneously—that is, as a space expands, it creates temporal particles. Thus, the universe maintains a constant energy level throughout its development. The creation of temporal particles in an excited volume of space and the consequent lowering of the energy of that space is counterintuitive in that most cases of particle creation are not associated with a decrease in the energy level of a system, but rather an increase. That is, usually when particles are created, the energy level of the system they are a part of is raised not lowered.

With the creation of temporal particles in the two spaces, the universe consisted of interior space, exterior space, and time—with t^+ existing in interior space and t^- existing in exterior space. The first moment actually occurred when a t^+ particle was created in interior space and a t^- particle was simultaneously created in exterior space, which is equivalent to t^0 transforming into t^+ (in interior space) and t^0 transforming into t^- (in exterior space). The operative word here is "equivalent," as there was no preexisting t^0. As noted earlier, each of these processes, t^0 to t^+ and t^0 to t^-, represents a single, fundamental moment of time. With all of the above concepts taken together, the Big Bang represents three simultaneous events in TET: 1) the temporal polarization of the primordial seed of green space—the development of a blue space and a yellow space, 2) the expansion of those two spaces—the expansion of the universe as a whole, 3) the creation of a t^+ particle in interior space and the creation of a t^- particle in exterior space—which occur simultaneously and represent the first moment of time.

Immediately following that first moment, interior space and exterior space begin to expand more. On the side of interior space, the additional expansion is a reaction of interior space to the t^- developing within it. That is, t^+ is created in expanding blue interior space, but owing to the blueness of that space,

---

* Some readers may consider this polarization to have occurred not spontaneously, but rather deliberately. This is fine and takes nothing away from the scientific discussion. It is interesting that upon polarization, especially once temporal particles were created within the blue and yellow spaces, the original universe, as a whole, resembled a particle of light from the TET perspective. The Judeo-Christian faith, as many know, also links the creation of the universe first to the creation of light.

it is quickly transformed into t^-, according to the t^+/t^- process. As t^- develops within interior space, it expands due to the repulsion it "feels" for the t^-. On the side of exterior space, the expansion is a reaction of exterior space to the t^+ developing within it, according to the t^-/t^+ process. As the two spaces continue to expand, more true-vacuum regions develop within them. Thus, more t^+ is created in interior space, and subsequently converted to t^-. And more t^- is created in exterior space, and subsequently converted to t^+— and so on.

As the expansion continues, interior space tries to eject its t^- into exterior space, while exterior space tries to eject its t^+ into interior space. Neither is able to accomplish this, as it is like two people in adjoining rooms trying to get to the opposite room through the same doorway at the same time. Each is blocking the other. With an inability to release the unwanted temporal energy and its continual build up, interior space and exterior space undergo a wild period of expansion. This instability causes one of them, in this case interior space, to "give in," which means turn green. (Which one turns green is random.) By turning green, interior space can tolerate its volume of t^- energy; there is no need for it to eject it anymore. Also, with interior space no longer trying to push its own temporal particles out the door, the doorway is free for exterior space to eject its t^+ into interior space. Having done this and being attracted to t^-, exterior space grabs all of the t^- energy still floating in interior space and expands wildly again as it converts it to t^+. This is because the volume of t^- exterior space grabbed was essentially the same volume that was causing interior space to expand wildly when it was blue. This second period of wild expansion settles down once exterior space ejects all of its extra t^+ into interior space.

These two periods of wild expansion are what TET considers to be cosmic inflation, the period following the Big Bang in which the universe is understood to have experienced rapid growth. TET, of course, has two such periods. In the first, both interior space and exterior space are expanding together. During the second period, exterior space pulls interior space along for the ride, which continues to occur today, although in a more calm and steady manner. From the standpoint of TET, the continued expansion occurs as exterior space converts the t^- it absorbs into t^+; exterior space expands due to the repulsion it feels for the t^+. The expansion does not become extreme because exterior space is able to eject its t^+ particles into interior space, although it does continue to sense the presence of t^+ particles even when they are in interior space. The physical acts of ejecting t^+ into interior space and pulling t^- in also cause exterior space to expand. All of this expansion, of course, means that temporal-particle creation continues to occur in both exterior space and interior space, maintaining the overall energy level of the universe. (With interior space now being green, both t^+ and t^- are created within that space as it expands, with t^- alone being created in exterior space as it expands.) The volume of each space increases exponentially (for example, from one unit to two to four to eight etc.), such that the number of temporal particles also increases exponentially, causing expansion to accelerate.

In TET, it was during the second period of inflation that the force carriers (in particular photons and Z bosons) and matter and anti-matter particles were created. This is because, during that period, interior space was green, allowing it to generate electric and weak field lines due to the tremendous fluctuations it experienced as it was pulled wildly by exterior space. (Strong field lines would come later, after the creation of quarks.) From the photons and Z bosons, matter and anti-matter arose. With t^+ dominating in interior space, electrons had an advantage over positrons. Some electrons and positrons annihilated each other to re-form photons, but with positrons not being able to process t^+ energy, some slowly suffocated temporally, turning into anti–electron neutrinos. The electrons, however, thrived and thus had greater numbers. Additional neutrinos and anti-neutrinos arose from Z bosons.

For quarks to have formed, photons would have to have separated in such a way as to create frames with mixed segments. That is, instead of creating frames with three blue segments and frames with three yellow segments, they would have to have given rise to frames with two blue segments and one yellow segment and frames with two yellow segments and one blue segment. Blue-blue-yellow and yellow-yellow-blue frames, however, are unstable, having starkly opposite segments within them. Normally, two identically polarized segments within a frame have the power to convert the third oppositely polarized segment to be like the other two, using the energy from the instability.

Thus, a blue-blue-yellow frame turns into a blue-blue-blue frame (an electron). However, with t^+ dominating in interior space, a yellow-yellow-blue frame does not transform into a yellow-yellow-yellow frame (a positron) because this would increase instability or, at best, exchange one type of instability for another: A yellow-yellow-blue frame is unstable because of the starkly opposite segments within it, but a yellow-yellow-yellow frame is also unstable because there is no segment within it that can breathe in the temporal (t^+) atmosphere of interior space. Thus, a yellow-yellow-blue frame transforms into a frame with two yellow segments and a bluish-green segment (an up quark). The blue segment has become more yellow, but not so much as to cause instability. It is also possible for a yellow-yellow-blue frame to transform into one with two yellowish-green segments and one blue segment (a down quark). The lone blue segment does not have enough power to pull both yellow segments fully into the blue range, but it can get them there a little, turning them yellowish green (*Figure 9.1*). The bluish-green segment in the up quark and the blue segment in the down quark help the yellow and yellowish-green segments in those particles survive; the yellow and yellowish-green segments use the t^- the blue and bluish-green segments give off.

Note that with blue-blue-yellow and yellow-yellow-blue frames changing as described, matter dominates over anti-matter in interior space, with the principal particles being electrons, up quarks, and down quarks. The domination of matter over anti-matter in the universe ultimately relates to time. Interior space is filled with t^+ particles rather than t^- particles, and this will affect how particles develop, as well as their stability. In our t^+-filled space, a blue-blue-yellow frame would more likely transform into an electron than an anti-up quark or an anti-down quark, making anti-protons and anti-neutrons rare. Positrons can of course be created directly from electromagnetic field lines, but they are unstable in interior space (except perhaps under special circumstances), as they are completely inca-pable of processing t^+. Positrons likely transformed into anti–electron neutri-nos, which are better able to exist in interior space, and as described above, yellow-yellow-blue frames likely transformed not into positrons, but up quarks and down quarks.

The separation of photons in such a way as to be composed of mixed segments also gives rise to the green segments in matter and anti-matter particles. For example, a frame consisting of three blue segments and one yellow segment

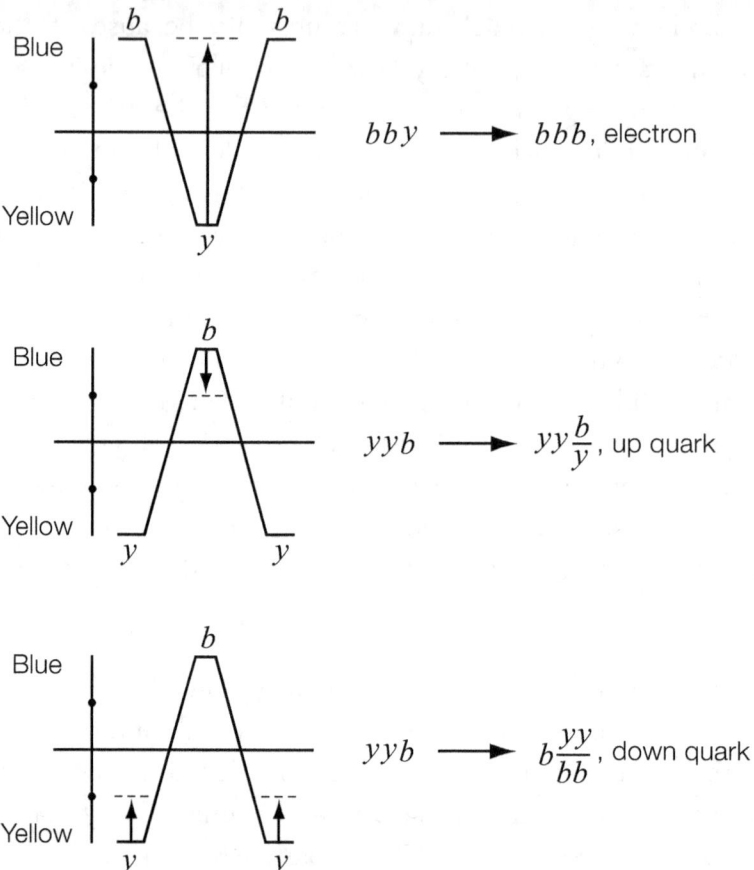

$$bby \longrightarrow bbb, \text{ electron}$$

$$yyb \longrightarrow yy\frac{b}{y}, \text{ up quark}$$

$$yyb \longrightarrow b\frac{yy}{bb}, \text{ down quark}$$

Figure 9.1 *When an electromagnetic field line separates in such a manner that a resulting particle has two blue segments and one yellow segment in its frame (an unstable situation), the two blue segments can use the energy of the instability to transform the yellow segment into a blue segment, producing an electron. The blue segments are more stable because they exist in the t^+-filled atmosphere of interior space. In the case of a frame that has two yellow segments and one blue segment, the blue segment becoming completely yellow would not lead to stability, as yellow segments cannot process the t^+ in interior space. Also, the blue segment does not have enough power on its own to raise both yellow segments fully into the blue range. Thus, one of two things can occur—either the blue segment becomes only partly yellow (bluish green), producing an up quark, or the yellow segments become partly blue (yellowish green), producing a down quark.*

would likely transform into a muon, and one with three blue segments and two yellow segments would likely turn into a tauon. The yellow segments in these particles turn green because there are already three polarized (blue) segments in the frames; no more polarized segments can exist, not even a yellowish-green segment. A frame with three yellow segments and one blue segment would likely transform into an anti-muon. Note that although the blue segment turned green, it still produces some t⁻, which aids the yellow segments. A frame with three yellow segments and one blue segment may also turn into a charm quark or a strange quark for the same reasons a yellow-yellow-blue frame would transform into an up quark or a down quark; the extra yellow segment in this case would become the green segment. A frame with three yellow segments and two blue segments may turn into an anti-tauon, top quark, anti-top quark, bottom quark, or anti-bottom quark. A frame with two blue and two yellow segments may turn into a charm quark, anti-charm quark, strange quark, or anti-strange quark (*Figure 9.2*).

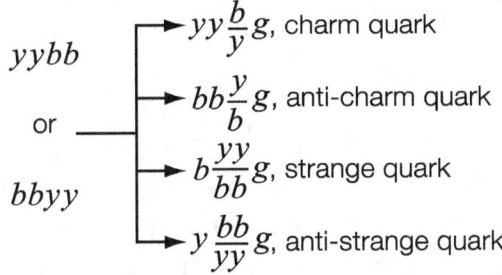

$bbby \longrightarrow bbbg$, muon

$bbbyy \longrightarrow bbbgg$, tauon

$yyyb \longrightarrow yyyg$, anti-muon; $yy\dfrac{b}{y}g$, charm quark; or $b\dfrac{yy}{bb}g$, strange quark

$yyybb \longrightarrow yyygg$, anti-tauon; $yy\dfrac{b}{y}gg$, top quark; $bb\dfrac{y}{b}gg$, anti-top quark;

$b\dfrac{yy}{bb}gg$, bottom quark; or $y\dfrac{bb}{yy}gg$, anti-bottom quark

$yybb$
or
$bbyy$

→ $yy\dfrac{b}{y}g$, charm quark

→ $bb\dfrac{y}{b}g$, anti-charm quark

→ $b\dfrac{yy}{bb}g$, strange quark

→ $y\dfrac{bb}{yy}g$, anti-strange quark

Figure 9.2 *Electromagnetic field lines may break apart in several ways, leading to the production of various types of matter and anti-matter particles.*

When photons separate in such a way as to give rise to particles with green segments, exotic particles may also be formed as byproducts, such as one with only two blue segments in its frame, having a fractional charge of –2/3. These particles quickly degrade. A particle with only two blue segments in its frame would have true-vacuum areas inside itself, as there would be areas of space too small to fit a temporal particle. The segments would likely collapse/merge into a single segment, with the energy transforming into electromagnetic radiation. Also, today, matter and anti-matter are created in more equal amounts than they likely were in the early universe. This is because today they are generated mostly from the field lines of preexisting matter and anti-matter particles, as opposed to general fluctuations in interior space, which are prone to produce electrons, up quarks, and down quarks, as described earlier. Preexisting matter and anti-matter particles, in a sense, influence the field lines to create particles like themselves and, as a consequence, sister particles.

At the end of the second period of inflation, the universe would likely have been filled with various elementary matter and anti-matter particles plus residual photons and Z bosons. Additionally, the quarks would have produced gluons. Muons, tauons, charm quarks, strange quarks, top quarks, bottom quarks, and all of their anti-matter counterparts would have quickly decayed due to their own instabilities. Continuing with the story briefly, up quarks and down quarks would join to create protons and neutrons, which would combine with electrons to form atoms, which would go on to form stars, planets, people, and everything else. The remaining photons, including Z bosons and gluons from the TET perspective, would be what is known as the cosmic microwave background radiation, the residual photons from the birth of the universe zooming around the cosmos even today.

Note that it would have been possible for both interior and exterior space to have turned green after the first period of inflation. If this had happened, both spaces, and thus the universe, would have collapsed. This would have occurred because green space is not repelled by either type of temporal particle. Green space also does not eject them, and a vacuum that is green will not, on its own, absorb them from its partner vacuum. It is these activities that prop up exterior space, preventing it from falling in on itself. The interior space of today, although green, does not collapse because it is being pulled by yellow exterior space as it expands. If both spaces turned green, they would collapse, absorbing their temporal particles along the way. The collapse would lead the universe back to its primordial state. A series of expansions and contractions—Big Bangs and Big Crunches—would continue to occur until only one of the vacuums turned green, leaving the other fully polarized.

Exterior space likely will not be able to expand forever. At a certain point, it may reach its maximum expansion level. This would not lead to a collapse, however, as exterior space would (1) still repel the t^+ it creates, (2) still be repelled by the t^+ particles it senses in interior space, and (3) continue to experience a push outward from its absorption of t^- and ejection of t^+. It simply would not expand further, developing into more of a "beating heart" universe, with slight expansions and contractions. Of course, with no further net expansion, no new temporal particles would be created in either of the spaces; a steady state would ensue.

Charge-Parity-Time Reversal Symmetry

The Charge-Parity-Time Reversal, or CPT, symmetry concerns the natural link between charge, parity, and time.[12] If you change one, you have to change the others in order for the laws of physics to operate the same. Specifically, CPT symmetry relates matter to anti-matter. For the universe to evolve with anti-matter in the same manner it has with matter, you would have to reverse the parity of the space and also the flow of time. The changing of the matter to anti-matter deals with the "charge" aspect of the theory. Reversing parity would change up to down, left to right, and top to bottom. To reverse time, of course, you make it run backwards. Note that the time referred to here is basic time—the time of elementary particles, not of composite structures, which is thermodynamic time.

CPT is directly related to TET's schematic of the universe (*Figure 9.3*). For ease, consider all matter to be represented by an electron and all anti-matter to be represented by a positron. If the electron were replaced by the positron, time and the parity of space would indeed have to be reversed for that

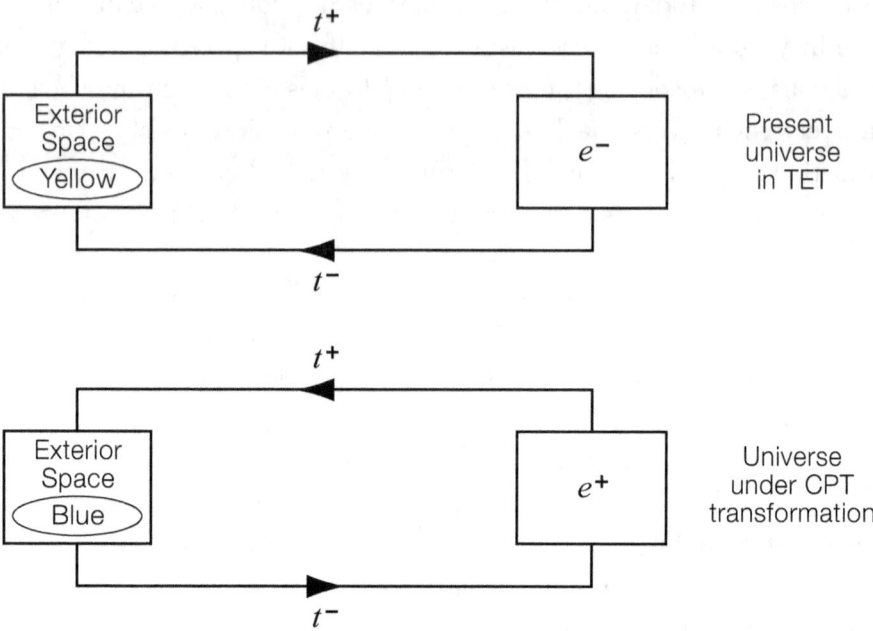

Figure 9.3 *TET's interpretation of CPT symmetry. For the universe to evolve in the same manner that it has currently, but with anti-matter (represented by e⁺) instead of matter (represented by e⁻), the flow of time must be reversed and exterior space must be blue instead of yellow. Interior space would still be green. Both spaces would have opposite coordinates from the universe in which matter dominated—that is, if a space had x, y, and z coordinates in the matter-dominated universe, it would have –x, –y, and –z coordinates in the anti-matter–dominated universe.*

positron to survive. The reversal of time is simple, in that positrons need t^- rather than t^+ to survive, so t^- would need to flow out of exterior space into interior space. Also, exterior space would need to be able to absorb t^+, the byproduct of the positron's temporal-respiration process. For exterior space to do this, it must be opposite what it is now—that is, it must be blue instead of yellow. The temporal polarization of a volume of space is an integral property of that space. If its other properties are switched, then its polarization will be switched. Thus, a change in parity is linked to a change in polarization. If up is now down, left is now right, and top is now bottom, the polarization, if yellow at first, will now be blue. Also, if the parity and polarization of one volume of space is changed, they must be changed in its sister space. Thus, up/down, left/right, and top/bottom would be changed in interior space, as well. However, with interior space being green, a change in polarization is a non-issue, as it would be like asking, "What is the opposite of zero?" That is, the opposite of green, in this instance, is green.

Note that the electron and positron were used above rather than "matter" and "anti-matter" more generally, because in TET, most anti-matter is capable of processing time in the t^+/t^- direction. But the positron, a very important anti-matter particle, is not. However, a positron-friendly universe, in which exterior space is blue and interior space is filled with t^- would evolve similarly to the universe we have now. In such a CPT-reversed universe, interior space would simply be filled with positrons, anti-protons, anti-neutrons, etc. Gravity and magnetism would operate via the t^-/t^+ mechanism, and the other forces would be the same as they are now.

Dark Matter and Dark Energy

For many years, scientists have known that 95% of the universe consists of some unknown forms of energy—referred to as dark matter and dark energy. As its name implies, dark matter is non-luminous, and it holds galaxies and galaxy clusters together. Dark energy is also non-luminous, and it plays a role in the expansion of the universe. Dark matter accounts for about 20% of the energy content of the universe, and dark energy accounts for about 75%, with ordinary matter accounting for about 5%. In TET, dark matter and dark energy are just temporal energy.

Recall that when space expands, temporal particles are created. Also, after their initial creation, they exist in pairs. That is, following the creation of the first t^+ particle in interior space and the first t^- particle in exterior space during the Big Bang, the particles come in multiples of two—their numbers would go from one to two to four to eight, etc. When exterior space expands, it creates only t^- particles within itself. When interior space expands, it creates one t^- particle for every t^+ particle. (If interior space were still blue, it would create only t^+ particles). To help understand dark matter and dark energy, consider a universe with room for two temporal particles in interior space and two temporal particles in exterior space, with all four eventually residing in interior space as t^+. For exterior space to maintain its polarization, only two of

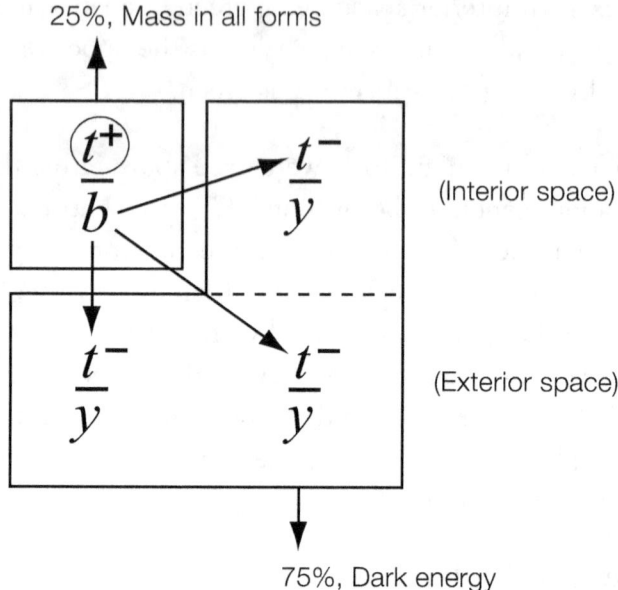

25%, Mass in all forms

(Interior space)

(Exterior space)

75%, Dark energy

Figure 9.4 *From the standpoint of TET, 25% of the energy content of the universe is available for mass in all forms, and 75% represents dark energy. The blue side of interior space is maintained through the continuous exchange of temporal energy between it and exterior space, as well as between it and the yellow side of interior space—in the figure, that energy is shown in its t⁻ state. This leaves one extra t⁺ particle (25% of the total energy) to serve as mass in all forms. The energy exchanged back and forth is dark energy.*

those particles must be bounced back and forth between interior space and exterior space, changing from t⁺ to t⁻ and back again accordingly. To maintain the yellow half of its nature, interior space must essentially toss another of those particles back and forth to itself. With the blue side of interior space itself being maintained through this activity, this leaves one t⁺ particle remaining (*Figure* 9.4). The three particles being bounced back and forth between interior space and exterior space and interior space and itself represent dark energy. As they are three of the four, they represent 75% of the energy content of the universe. As much of this energy goes to maintaining the polarization of exterior space, it also plays a role in the expansion of the universe.

The other temporal particle represents all matter, including dark matter. As it is one of the four, it represents 25% of the energy content of the universe.

This temporal particle is available to serve as mass. In the present universe, of course, there are more than four temporal particles. A vast amount of temporal particles exists to serve as ordinary matter, dark matter, and dark energy, but the percentages remain the same. Unlike ordinary matter, however, in which mass is that volume of temporal particles associated with the frame of a particle, the mass of dark matter is more "free." In some cases, dark matter is just random accumulations of temporal particles in space. Sometimes, these accumulations are seeded with neutrinos of various types or other particles. Also, the temporal particles of dark matter will surround ordinary matter, such as that in planets, stars, galaxies, and galaxy clusters, helping to hold them together.

It is important to understand that dark matter is for the most part just accumulations of temporal particles, which of course is just time. From afar, such an accumulation, particularly with the spatial compression that goes along with it, will look and behave like a normal mass, but if you were right on top of it, indeed even within it, you would notice nothing out of the ordinary—nothing more than what you would call empty space.

The Higgs Field

The Higgs field is the theoretical field in Quantum Mechanics that bestows mass on particles like electrons. From the standpoint of TET, this field, like the time dimension of General Relativity, is an approximation of the entire field of temporal particles existing in interior space—the gravitational field proper. Although only 25% of this field is available for mass, matter (and antimatter) can theoretically be created anywhere in space. A temporal particle that was once dark energy could become part of a mass field, and a temporal particle that was once a part of dark matter could become a part of dark energy. Temporal energy is always shifting about. This is likely the reason that the values of dark matter, dark energy, and ordinary matter are sometimes said to be about 23%, 73%, and 4%, respectively.[13] The ordinary matter, in a sense, has one foot in dark matter and one foot in dark energy. Higgs bosons, quanta of the Higgs field, may be like dark matter, general accumulations of temporal energy, but on a small scale, although likely not seeded with such particles as neutrinos. Supporting this idea is the fact that the Higgs field of QM and gravitational field proper of TET are both based on tachyons, faster-than-light particles—that is, temporal particles are tachyons.[14] When these particles accumulate, or condense, they bestow real mass on objects.

Black Holes

A black hole is a region of space with such strong gravity that nothing, not even light, can escape from it. It may be created through the collapse of a star. As the matter of the star is crushed into a smaller and smaller volume of space, the strength of gravity in the area rises. The matter of the star and anything that enters the black hole, after passing a point of no return called the event horizon, are destroyed by the intense gravity. A large misconception about black holes is that all of the matter that went into the black hole remains, just reduced in size. Actually, it is obliterated, leaving the black hole as little more than space and time.[15]

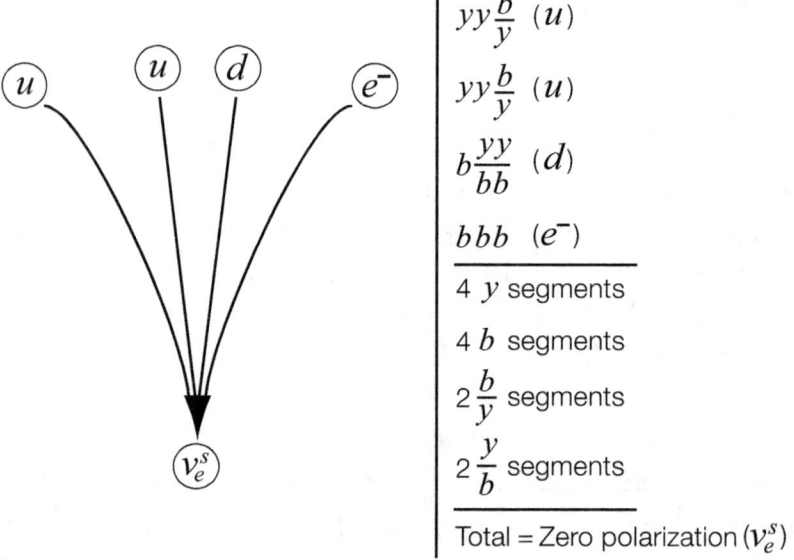

Figure 9.5 *From the perspective of TET, the frames of all the elementary particles going into a black hole overlap due to the compression. If any oppositely charged electrical or weak segments overlap, their effects will cancel. In the example shown, two up quarks, one down quark, and one electron merge to form a sterile neutrino.*

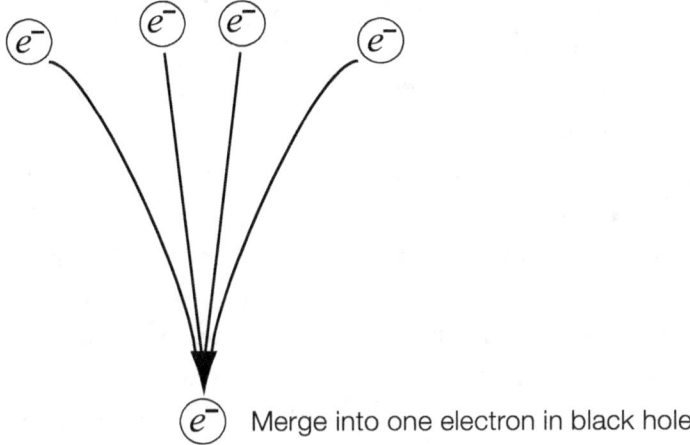

Merge into one electron in black hole

Figure 9.6 *If several electrons were compressed into one, the final frame would have the charge of just a single electron.*

From the perspective of TET, the frames of all the elementary particles going into a black hole overlap due to the compression. Consider a hydrogen atom, which is composed of one proton and one electron. As it falls into the black hole, the frames of the two up quarks, one down quark, and one electron composing the atom are fused into one frame. Recall that a particle frame is nothing more than space, perhaps polarized space. There is no limit to how much space can overlap on itself, except the point at which the primordial state of the universe is reestablished. Note also that there is no thickness to these overlapping areas of space. In the case of the hydrogen atom, the polarizations of the various segments in the overlapping particle frames cancel each other out, such that the end result is a sterile neutrino (*Figure 9.5*).

In TET, all the matter of a black hole is compressed into a single particle frame; this frame may have electric or weak charge or no charge, depending on the matter that went into it. If a frame has charge, it would be that associated with a single particle—that is, charge does not add up. For example, if several electrons were compressed into one, the final frame would have the charge of just a single electron (*Figure 9.6*). This is because, regardless of the overlap, the temporal respiration that occurs on the surface of this final frame would be enough to sustain just one electron. The other frames, lying in a sense underneath, would suffocate, turning green as a result. These "buried" frames would fade into the background from which they arose, with the final frame being the sole survivor. Because the other frames fade back into the vacuum from which they arose, there is no real loss of "information." That is, the energy still exists, just in a different form. It is not unlike an ice cube melting into a liquid volume of water in a glass. The molecules that composed the ice cube still exist, but now in a liquid form rather than a frozen form.

Note that there is a sense in which the general loss of t^- through singularities (in black holes or other parts of space) can be viewed as the loss of information—as if information mysteriously vanished from our space. However, this apparent loss of information is canceled when you consider exterior space and interior space together. The information that vanished from interior space appears in exterior space, such that when you view the universe as a whole, no information is actually lost.

Because most matter exists as neutral matter, with equal amounts of protons and electrons, most final frames should be that of sterile neutrinos. Also, the spins of all the particles going into a black hole would likely cancel each other for the most part, leaving at best a spin that moved in random directions, which is also characteristic of a sterile neutrino. The greenness of sterile neutrinos (corresponding to an appearance of being massless) and their random spins make them blend into the vacuum very well, particularly if they are not moving through space at high speed, as is the case when they are at the center of black holes. The result is that, even from the TET perspective, a black hole would appear to be nothing more than space and time. However, the sterile neutrino would still be there and continue to process time in the t^+/t^- direction.

Because of the immense amount of temporal energy in a black hole, any final frame would likely be compressed to its limit. The smallest a matter (or anti-matter) particle can be and still be a viable particle is just slightly larger than a temporal particle. Indeed, at its smallest, the diameter of an elementary matter particle like an electron or sterile neutrino, is nearly the same diameter as a temporal particle, approximately 10^{-35} meters. When such a particle is

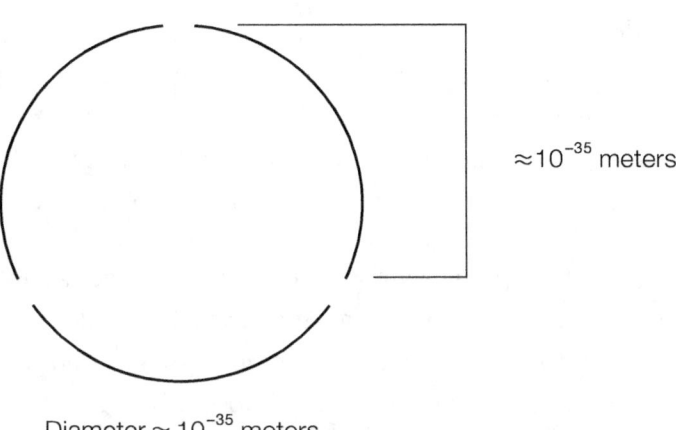

Diameter ≈ 10^{-35} meters

Figure 9.7 *When a particle with three segments in its frame is compressed to such an extent that its diameter is approximately 10^{-35} meters, each segment of the frame is also about 10^{-35} meters long.*

this size, the three segments composing its frame are also each compressed to a length of approximately 10^{-35} meters, which is also the length of a singularity. Thus, with temporal respiration occurring on each of the three segments there would be three singularities around the particle, through which its t^- particles are removed (*Figure 9.7*).

It has been known for some time that black holes are not necessarily permanent structures—that they may evaporate.[16] In brief, the mechanism by which this occurs has been described as follows: As a black hole forms, the space just outside of the event horizon fluctuates to such an extent that it produces particle/anti-particle pairs, with one particle of a pair having positive energy and its partner having negative energy. The particles may be of any QM type: for example, electrons and positrons. If the particle with positive energy is radiated away and its negative-energy sister falls into the black hole, the negative energy of that particle will eat away at the positive energy of the black hole and thus cause it to slightly shrink or evaporate. If this continued to happen, the black hole could completely evaporate.

In TET, the radiation of particles just outside of the event horizon deflects some temporal energy from entering the black hole, as that temporal energy responds instead to the energy radiating away. With the reduction in the amount of temporal energy flowing into the black hole and the continued consumption of the temporal energy already within it, the black hole slowly shrinks. From the TET perspective, the concept of particles with negative energy entering the black hole and consuming it is just a reflection of the TET concept of t^+ particles being transformed into t^- particles, and t^- particles being absorbed by exterior space.

Note, however, that if a particle with negative energy (specifically negative mass, t^- energy) was indeed the one to fall into the black hole, as opposed to its positive-energy partner, the black hole would evaporate more quickly. This is because, in this case, the energy flowing into the black hole is already in the form of t^-. As such, it does not add to the positive energy of the black hole and is readily available for absorption by exterior space. When a particle whose mass is mostly t^+ falls in, that t^+ must first be converted to t^- before it can vanish, causing evaporation to occur more slowly.

Blue Shifting and Red Shifting of Electromagnetic Waves

The energy of light, an electromagnetic wave, is usually described in terms of how "blue" or "red" the light is. The light may not literally be these colors. Blue light has a great deal of energy and is associated with a short wavelength. Red light, however, has a lower amount of energy and is associated with a longer wavelength—blue and red are on opposite sides of the visible portion of the electromagnetic spectrum. The blue shifting of an electromagnetic wave concerns any general shift toward the high-energy end of the electro-magnetic spectrum (as a whole) and is associated with a decrease in wave-length. Red shifting concerns any general shift toward the low-energy end of the spectrum and is associated with an increase in wavelength.

Consider two light-emitting objects located an equal distance from a stationary observer. Both move from their shared location at an equal speed. However, object A moves toward the observer, and object B moves away from the observer. The light emitted by object A will appear to have a shorter wave-length than the light emitted by object B. That is, object's A light will appear more blue, whereas object B's light will appear more red.

In TET, the reason this occurs is because the light emitted toward the observer by object A is compressed more than the light emitted toward the observer by object B. Recall the TET idea that temporal particles converge on moving objects, but move in principally from the front, in direct line of an object's motion, meeting it head on. Thus, there is more spatial compression in front of a moving object than there is behind it. Therefore, object A's light will be more compressed than object B's light. Object A is moving toward the observer, so he sees the more compressed end of that object and the light it is emitting. Object B is moving away from the observer, so he sees the less compressed end of that object and the light it is emitting. The more compressed end of object B and its compressed light are on the other side, out of view of the observer.

When an electromagnetic wave is compressed, as is the case for object A's light, there are fewer temporal spaces along its frame to process temporal particles. Therefore, it carries more of these particles as mass, which equates to higher energy, which in turn equates to a shorter wavelength or bluer light. When an electromagnetic wave is more expanded, as is the case for object B's light, there are more temporal spaces along its frame to process temporal particles. Therefore, it carries less of these particles as mass, which equates to a lower energy, which in turn equates to a longer wavelength or redder light.

Note that when both waves leave their respective areas of compression, they will each expand to some degree, or red shift; however, object A's light will be less red shifted than object B's light because object A's light carries more energy. Thus, object A's light would still be measured to be more blue than object B's. And as object A approaches the observer, its light will appear to get even more blue, as the area of compression is getting closer to the observer. As object B moves further away from the observer, its light will appear to get more red, as its light travels into areas with less and less spatial compression the more its source moves away.

The expansion of space also causes the red shifting of light. As space expands, it also expands electromagnetic waves, causing the same effects mentioned above—an increase in the number of temporal spaces along the wave with a concomitant decrease in the amount of temporal energy carried as mass. Light also red shifts as it, for example, leaves the sun. This is because the space surrounding the sun is more compact toward its surface. As the light leaves the surface, it enters into more expanded areas of space and thus itself expands, red shifting as a result.

Note some have attributed the loss of energy due to red shifting simply to the stretching of the wave. However, as stretching a rubber band does not reduce the amount of mass in the band, stretching a light wave also does not, in and of itself, lower the energy of the wave. The loss of the wave's energy is attributed in TET to an increase in the number of temporal spaces along the wave's frame as it expands. With an increase in the number of spaces, more temporal energy is processed as time rather than carried as mass/energy—see *Figure 5.4* for an analogous situation involving matter.

PART

3

SOME ADDITIONAL DETAILS

CHAPTER
10

SPACE, TIME, AND ENERGY

Although this book steers away from mathematics, there are a few very basic mathematical ideas that may help in further understanding TET, relating to space, time, and energy. They deal with some fundamentals of size, mass, and time; the equivalence of matter and energy; and the discrepancy between the measured and calculated values of the vacuum energy—known as the vacuum energy problem. They are discussed here only generally.

Fundamentals of Size, Mass, and Time

The smallest length that is said to have any physical meaning is 1.616×10^{-35} meters, known as the Planck length.[17] The reason for this in TET is because it is the diameter of a temporal particle (Figure 10.1), which is likely only a two-dimensional object that exists most comfortably in a volume of space at least one cubic Planck length. (Temporal particles are always jittering about, and such a volume would allow it room enough to rotate 360 degrees in any direction.) Any length of space smaller than a Planck length would not be able to interact with a temporal particle.

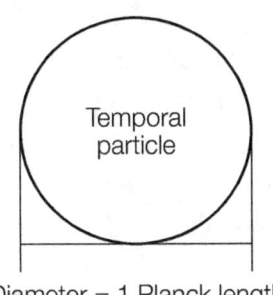

Diameter = 1 Planck length

Figure 10.1 *A temporal particle is considered to have a diameter of 1.616×10^{-35} meters, equal to 1 Planck length.*

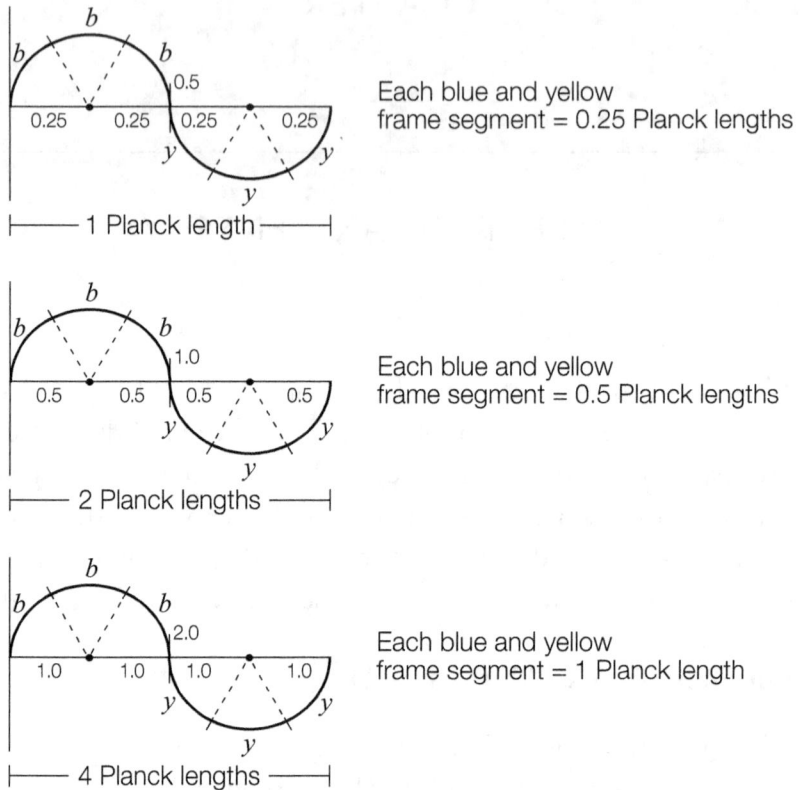

Figure 10.2 *Three electromagnetic waves. Only the third wave can fully process time along its entire length, as each of its segments is 1 Planck length long, about the size of a temporal particle. The wave is capable of sustaining its full electromagnetic nature.*

Now, consider three electromagnetic waves: with the first being 1 Planck length long, the second being 2 Planck lengths long, and the last being 4 Planck lengths long (*Figure 10.2*). The first wave is incapable of sustaining an electromagnetic character because neither the blue wave nor the yellow wave contains a full Planck length—they each contain three quarters of a Planck length. Thus, temporal respiration cannot occur along the length of the electromagnetic wave, and its blue and yellow natures would begin to fade to green. The inability of the blue and yellow waves to process time also makes the whole wave unstable. However, the wave would still be unstable and unable to process time even upon losing its blue and yellow natures. This is because, despite having a full, green Planck length at that point, a temporal particle would have difficulty interacting with it due to the up/down orientation of the half waves.

The second wave is capable of sustaining some measure of an electromagnetic character. This is because both its blue and yellow waves contain a full Planck length. However, they each also contain a segment one half of a Planck length. In each wave, this half Planck length would fade to green. Even though they represent one full Planck length together, one is blue and the other yellow; as such, they would be incompatible as a single polarized segment. However, the 2 Planck-length wave would still be unstable even upon the two half Planck-length segments fading to green. This is because even though the two halves are able to function as a single segment after turning green—processing t^+ sometimes and t^- other times—that green segment is mixed with fully polarized segments. Recall from chapter 3, that a green segment mixed with fully polarized segments leads to instability, as in the case of a muon or tauon.

The third electromagnetic wave, which is 4 Planck lengths long, can fully process time along its entire length and thus is stable and able to sustain its whole electromagnetic nature, as its blue and yellow waves are composed of three full Planck lengths each. In TET, all fully electromagnetic waves are expansions of the wave 4 Planck lengths long. As a consequence, the energy of every other electromagnetic wave is a dilution of the energy of this wave— as this wave has the highest concentration of energy a fully electromagnetic wave can have. The energy of the wave is diluted when it is stretched to longer wavelengths. When the wave is stretched, more temporal spaces are available along its length or frame, such that more of its temporal energy can be, and is, processed as time rather than carried as mass (see the discussion on red shifting in the previous chapter).

In TET, each Planck length in the blue and yellow waves of an electromagnetic wave 4 Planck lengths long is associated with a mass of approximately 2.17×10^{-8} kilograms (kg) (as an absolute value, not positive or negative), called the Planck mass (*Figure 10.3*)[17]; it is the greatest mass a fully polarized line of space this size can have, although this does not mean that every line

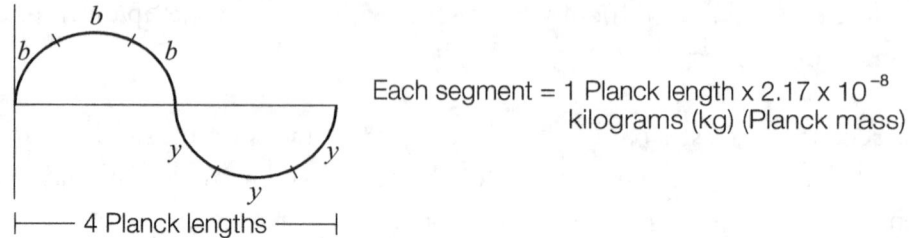

Each segment = 1 Planck length x 2.17 x 10^{-8} kilograms (kg) (Planck mass)

Figure 10.3 *In TET, all fully electromagnetic waves are expansions of the wave 4 Planck lengths long. This wave has the highest concentration of energy a fully electromagnetic wave can have. The energy of every other electromagnetic wave is a dilution of the energy of this wave.*

of space this size has the maximum mass. For example, a Planck-length segment in an electromagnetic wave 10^{-6} meters long has a mass much less than 2.17 x 10^{-8} kg. As a whole, the electromagnetic wave 4 Planck lengths long has a mass of 3.4 x 10^{-8} kg. This is roughly equivalent to 2.17 x 10^{-8} kg x 6 (for the number of blue and yellow segments) divided by 4 (for the length of the wave as a whole).

This wave is capable of giving rise to an electron and positron, each with a mass of 3.4 x 10^{-8} kg (*Figure 10.4*). Each particle has the same mass as the whole wave because as the blue and yellow waves separate and wrap back on themselves, their area is reduced by half, which doubles their energy. That is, the electron has double the energy of the blue wave, and the positron has double the energy of the yellow wave, making each of them equal the energy of the full wave. The diameter of the electron and positron in this case is 1 Planck length. An electron and positron of this size represent the smallest these particles can be and still be an electron and positron. When an electron and positron are compressed to this size, each of the three segments in each of their frames is also 1 Planck length long. Their Planck-length size allows the segments to interact with temporal particles to maintain their polarization— and these three segments are the origin of electric charge being divisible into one-third units. If the frames were compressed further, the segments would be too small to interact with temporal particles fully, and they would begin to temporally suffocate, losing their polarization. Also, their internal volumes would be too small for a temporal particle to fit inside, leading to a true vacuum

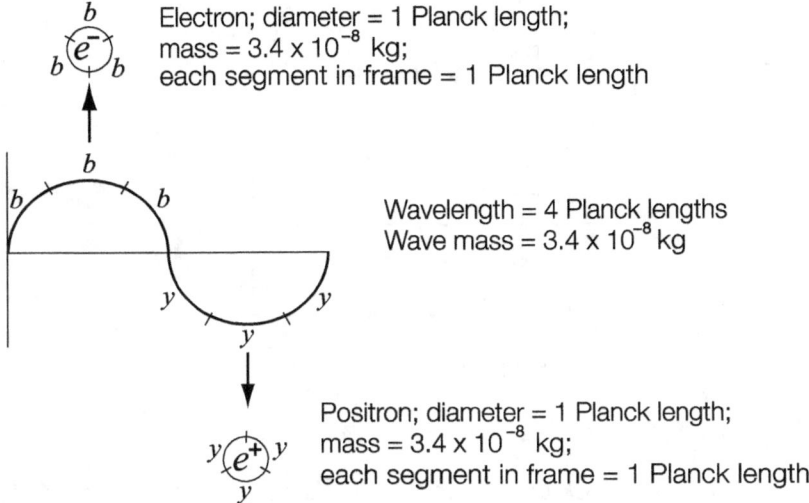

Electron; diameter = 1 Planck length;
mass = 3.4 x 10^{-8} kg;
each segment in frame = 1 Planck length

Wavelength = 4 Planck lengths
Wave mass = 3.4 x 10^{-8} kg

Positron; diameter = 1 Planck length;
mass = 3.4 x 10^{-8} kg;
each segment in frame = 1 Planck length

Figure 10.4 *An electromagnetic wave 4 Planck lengths long is capable of giving rise to an electron and positron, each with a mass of 3.4 x 10⁻⁸ kg.*

developing within each of them. Any volume of space devoid of temporal particles is highly excited, leading to instability within that space.

Note that the wave that would actually start the process of creating the electron and positron—the "incident" wave—would be the wave 2 Planck lengths long. Incident waves are half the size of the waves that, in TET, are considered to actually give rise to matter/anti-matter pairs. Being half the size, incident waves have double the energy of the later waves, and thus have the same mass as the matter and anti-matter particles combined. TET considers an incident wave to transform into a longer wave from which the matter and anti-matter particles actually arise. With the wave 2 Planck lengths long being only partly electromagnetic, some of the energy needed to create the fully electromagnetic 4 Planck length wave and, ultimately, the electron and positron would come from the energy of its instability.

The Planck time is the fundamental moment of time. It has a value of 5.39 x 10^{-44} seconds.[17] In TET, it represents the transformation of t$^+$ to t^0, t^0 to t$^-$, t$^-$ to t^0, or t^0 to t$^+$. Another, equivalent way of describing it is that it is the time it takes for light to traverse a distance of 1 Planck length, which

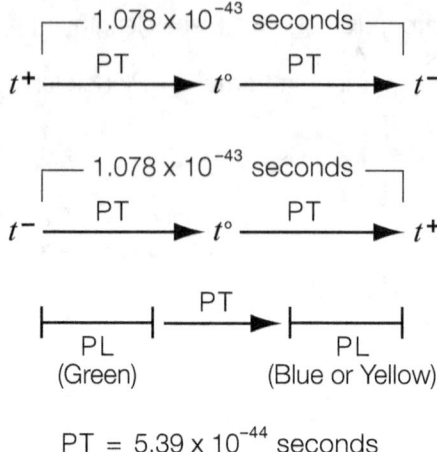

$$PT = 5.39 \times 10^{-44} \text{ seconds}$$

Figure 10.5 *In TET, the Planck time is the time it takes for a temporal particle to transform into a different state, for example t^+ to t^0, or for a Planck-length (PL) segment of space to become temporally polarized. It has a value of 5.39×10^{-44} seconds. The time it takes for t^+ to transform into t^-, or vice versa, is 1.078×10^{-43} seconds, or twice the Planck time.*

in TET simply means the time it takes for a Planck-length segment of space to become temporally polarized. Thus, the Planck time is the time it takes for particle time (t^+ or t^-) to become spacelike (t^0), or for space (green) to become timelike (blue or yellow), with the transformation of t^+ to t^0, t^0 to t^-, t^- to t^0, and t^0 to t^+, as well as of a Planck-length segment of green space to blue or yellow space, each representing the speed of light (approximately 3×10^8 meters/second in vacuum),[17] as described earlier. Another time that is considered to have significance in cosmology is 1.078×10^{-43} seconds, which is twice the Planck time. In TET, it represents the full process of t^+ to t^- or t^- to t^+—that is, each full process contains two subprocesses of one unit of Planck time each (*Figure 10.5*).

Equivalence of Matter and Energy

One equation that even many young children know is $E=mc^2$. This was discovered by Albert Einstein and says that energy is equivalent to matter (or anti-matter) by a factor of the speed of light squared, designated as c^2. Through this formula, energy can be converted to matter, and matter can be converted to energy. TET uses this equation but breaks it down into space and time. In TET, the "E" is spatial energy, and the "m" is temporal energy in the form of mass. What $E=mc^2$ is saying from the perspective of TET is that spatial energy is equivalent to temporal energy by a factor of the speed of light squared. In the theory, one "c" corresponds to the conversion of t^+ to t^0, and the other "c" corresponds to the conversion of t^0 to t^-, with the original t^+ coming from the mass field. (It works the same if the mass is composed of t^- particles, just in reverse.) Basically, the events involved in the conversion of mass energy (time) from t^+ to t^0 and t^0 to t^- bestow energy on the particle frame (space) (*Figure 10.6*).

As noted above, the equation has been used to understand the conversion of energy to matter, and matter to energy. The energy referred to in this instance has typically been radiant energy in the form of light, although Z bosons and gluons are also possible. Therefore, the equation has been used, and successfully, to indicate that radiant energy can be converted into matter and matter into radiant energy. The reason this operation has been successful from the

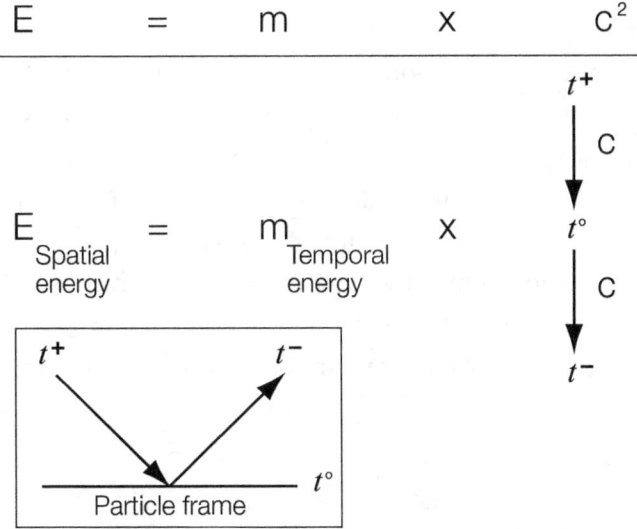

Figure 10.6 *In TET, E = mc² represents spatial energy equaling temporal energy (as mass) times the speed of light (c) squared, in which one c represents the transformation of t^+ to t^0 and the other represents the transformation of t^0 to t^-. The attraction of t^+ to the particle frame, the conversion of t^+ to t^-, and the repulsion of t^- all bestow energy on the particle frame.*

standpoint of TET is because it is essentially comparing apples to apples. In TET, matter and radiant energy are composed of the same stuff—temporally polarized space and temporal particles. Thus, making "E" represent radiant energy and "mc^2" represent matter energy is equivalent to making "temporally polarized space and temporal particles" equal "temporally polarized space and temporal particles." Thus, you can use the equation to track the conversion of matter into radiation and vice versa.

The "E" sometimes also represents kinetic energy—a form of energy associated with motion. This is also equivalent because kinetic energy is also composed of space and temporal particles. The object moving is space, typically a particle frame, but really any space. Associated with this moving space are temporal particles, which are attracted to the motion energy. Thus, the moving object, its mass, and the temporal particles responding to its motion energy together represent kinetic energy.

The Gravitational Constant

The gravitational constant (G, pronounced "big G") is an important value, and it is worth exploring its relation to TET concepts. In TET, the constant represents the destruction of a mass field, which allows gravity to occur. That is, gravity occurs when t^+ is converted to t^- and this is subsequently absorbed by exterior space. As this transformation occurs via a particle's frame, gravity occurs principally through the elimination of the temporal particles specifically responding to the energy of the frame, some of which are directly bound to the frame. That is, gravity occurs through a particle's mass. Of course, mass is continuously replenished as more temporal particles move in, but its continuous destruction is the "engine" of gravity.

Consider two electrons (A and B) interacting through gravity only. Let us assume they are compressed to the extent that each is 1 Planck length in diameter (*Figure 10.7*). The mass of each electron thus becomes 3.4×10^{-8} kg—up from 9.1×10^{-31} kg when it has a diameter of about 6×10^{-13} meters. With half of one particle facing half of the other, the combined mass in the middle is also 3.4×10^{-8} kg. For gravity to occur, this common mass has to be

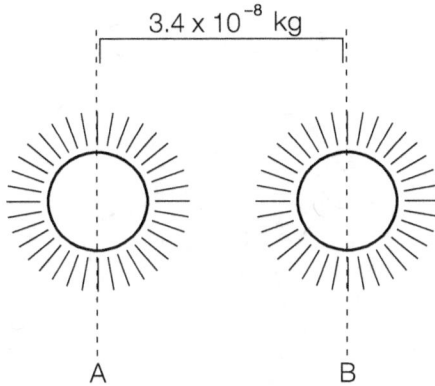

Figure 10.7 *Two gravitating electrons, ignoring their electric repulsion. In this example, the electrons are each 1 Planck length in diameter and have a mass of 3.4 x 10⁻⁸ kg. The total mass between them, on their gravitating sides, is also 3.4 x 10⁻⁸ kg.*

destroyed, which will occur essentially through the particles' frames. The temporal particles in the common mass field will be processed by either electron A's frame or electron B's frame, such that on average, all of the temporal particles are processed by one frame—actually one half of one frame. Another way of viewing this is to distill the common mass field down to one temporal particle at a time; that particle will be processed by electron A or B but not both. Let us assume it is processed by electron A, then assume every other temporal particle is processed by electron A. Gravity will have occurred because the common mass field will have been destroyed. It does not matter which frame processed the particles, and it is indeed easier mathematically to regard all of the common mass to be processed by just one of the particles.

The equation in *Figure 10.8* demonstrates the destruction of a mass field. The "1" in the value 1.57 represents one full Planck-length-long polarized segment. The 0.5 represents the half Planck-length-long polarized segment in the half frame of the electron. It is fine to use a half Planck length in this instance because it comes from a full Planck length—the division of that Planck length is imaginary. However, only half of the effects of temporal respiration on that Planck length play a part in gravitational attraction on the side of the electron. The 0.07 comes from the fact that the gaps between the

$$G = \frac{1.57\ PL}{3.4 \times 10^{-8}\ kg} \times c^2 \approx 6.7 \times 10^{-11}\ \text{meters}^3 / \text{kg-seconds}^2$$

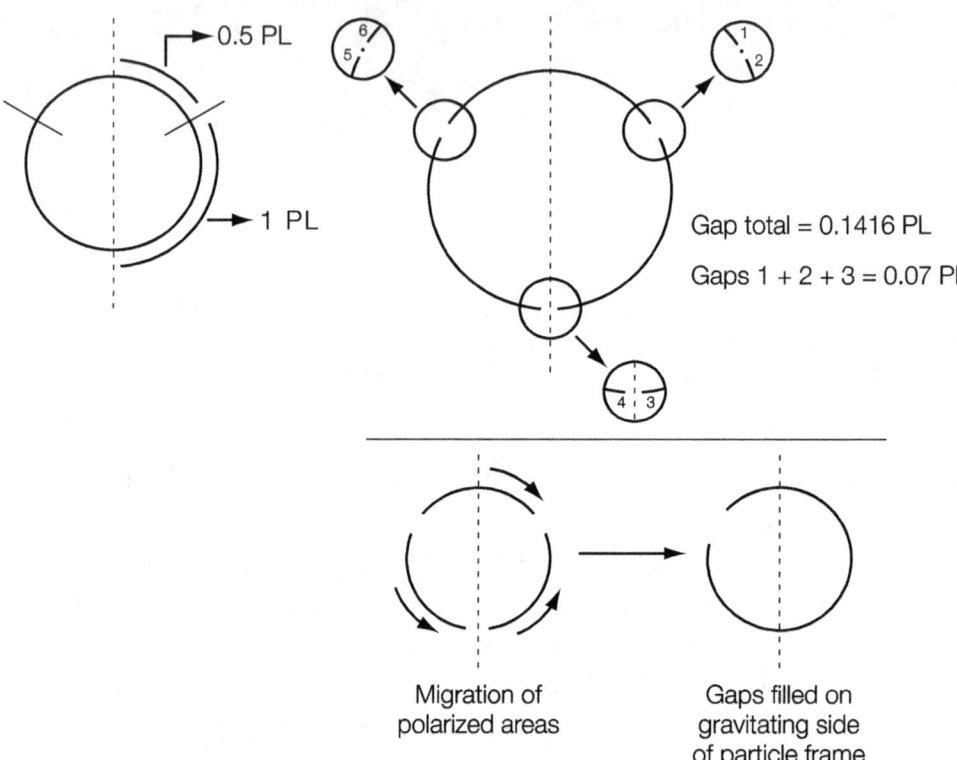

Figure 10.8 *In TET, G represents the destruction of the mass between two gravitating bodies, which in turn represents the "engine" of gravity. The mass can be considered to be consumed by only one of the frames involved in the interaction—actually half of that frame—by the speed of light squared. Using the two gravitating electrons from Figure 10.7, the mass to be consumed is 3.4 × 10⁻⁸ kg. Half of either electron's frame has a length of 1.57 Planck lengths. Although only 1.5 polarized segments officially exist in half the frame, all of the polarized areas along the frame can migrate, filling the gaps on the gravitating side, which gives the effect of there being a 1.57 Planck-length-long polarized segment along the half frame.*

three polarized segments add up to 0.1416 Planck lengths. (This is determined from pi, which is 3.1416, minus 3 for the 3 Planck lengths present.) Sometimes the polarized areas within a frame migrate, filling those spaces, with the effects of their temporal respiration playing a part in gravitational attraction.

If you consider each gap to represent two spaces that can be closed up, there are six total along the frame. As shown in *Figure 10.8*, half of the electron's frame would consist of three of those spaces; thus 0.1416 Planck lengths divided by six times three, or 0.07 Planck lengths, would play a role in gravity on that side of the electron just as the 0.5 segment does. Note that although they are represented as such, the gaps are not actually openings in the frame; they are gaps in polarization only. They are small areas within the frame without polarization. Because of their small size and because they can change locations along the frame, they do not represent hot spots and do not cause instability in the particle.

Adding everything up, you get gravitational effects from 1 plus 0.5 plus 0.07 Planck lengths, with the value of G coming out to be 6.7×10^{-11} meters^3/kg-seconds^2, which is close to the experimental value of 6.67×10^{-11} meters^3/kg-seconds^2. The full equation for the force of gravity of course involves G, the masses of the particles gravitating, and their distance from one another (*Figure 10.9*).

$$F = G \frac{m_1 \, m_2}{r^2}$$

(Law of Gravitation)

Figure 10.9 Newton's Law of Gravitation. G is the gravitational constant. M_1 and M_2 represent the mass of the gravitating bodies, and r^2 represents the distance between the bodies squared.

Thinking of the 1.57 Planck lengths as one whole, the division of it by the 3.4×10^{-8} kg of mass in the determination of G is analogous to a single doorway being divided by, for example, 10 people. The temporal particles of the mass surround the 1.57 Planck lengths like the 10 people surround the doorway. The temporal particles, in a sense, go through the 1.57 Planck lengths to enter exterior space, like the people go through the doorway to enter perhaps an adjacent room. Of course, temporal particles enter into full Planck lengths of space. However, the effect on the gravitating side of the particle is that the 3.4×10^{-8} kg of mass are going into 1.57 Planck lengths of space. Note that the value 1.57 Planck lengths—rather than 1.57 Planck lengths cubed (for a Planck-scale volume of space)—is used because although temporal particles live in whole volumes of space, they leave interior space for exterior space through one dimension only. It does not matter if that dimension is considered to be the x, y, or z dimension.

The speed of light squared (c^2) in this instance relates to the transformation of t^0 to t^- (one "c") and t^- to t^0 (the other "c"). The first process represents energy leaving the mass field (as t^0) and entering the gravitational field (as t^-). The second process represents energy leaving the gravitational field (as t^-) and entering exterior space (as t^0), upon which full gravity can be said to have occurred *Figure 10.10*.

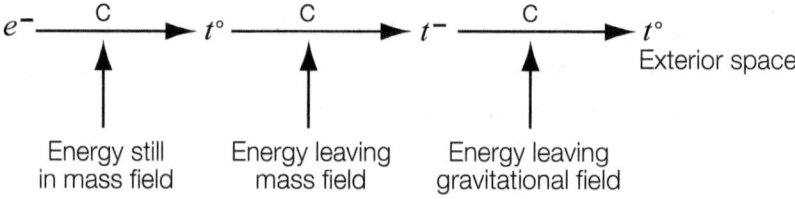

Figure 10.10 *Representation of c^2 in gravity, from TET perspective. The first c does not play a role, as it is associated with energy still within the mass field. In TET, gravity involves energy leaving the mass and gravitational fields.*

Although the example above deals with two electrons of equal size and the Planck length in diameter, the value of G would be the same for electrons of different sizes, which would be due to their being in volumes of space with different degrees of compression. A scale factor would simply need to be applied to the mathematics. Also, as electrically neutral matter is typically a positive nucleus surrounded by electrons, the continuous consumption of electron mass drives many instances of gravitational attraction—this is another view of the electric force being the foundation of gravity.

The Coulomb Constant

The equation describing the force of electric attraction or repulsion is similar in form to that describing gravity (*Figure 10.11*). As G represents the engine of gravity, K represents the engine of the electric force. In TET, K rep-

$$F = K \frac{q_1 \, q_2}{r^2}$$

(Coulomb's Law)

Figure 10.11 *Coulomb's Law for electric attraction and repulsion. K is the Coulomb constant. q_1 and q_2 represent the charge magnitude of particles 1 and 2, and r^2 represents the distance between the particles squared.*

resents the interaction between the initial primary lines emanating from the interacting particles. *Figure 10.12* shows TET's description of K. The expression in the numerator represents the two primary lines. There are three segments in each of those lines—each of the segments is represented here as the small-

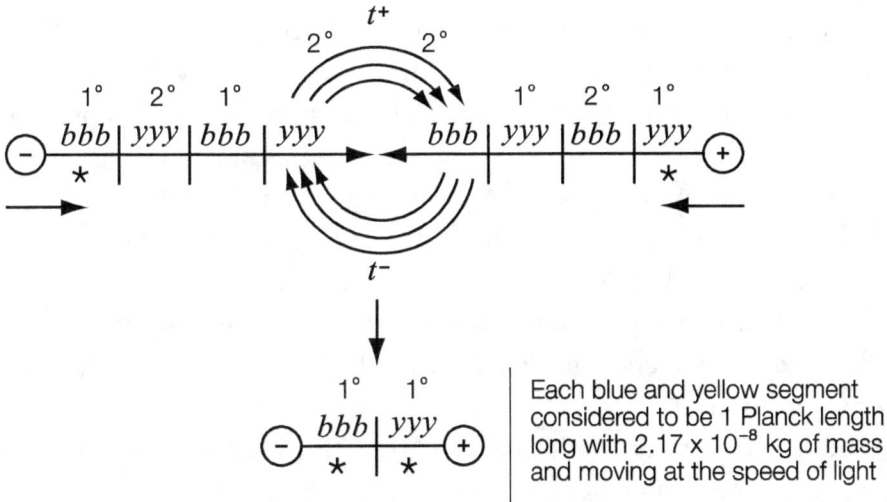

$$K = \frac{(2)(3 \times PL \times 2.17 \times 10^{-8} \text{ kg} \times c)}{(1.60 \times 10^{-19} C)^2 \times \ (6 \times 137c)} \times c^2 \approx 8.99 \times 10^9 \ \frac{\text{kg} \cdot \text{meters}^3}{C^2 \cdot \text{seconds}^2}$$

Figure 10.12 *K represents the "engine" of the electric force. In TET, it concerns the interaction between the initial primary lines emanating from the interacting particles. In the equation, the two primary lines are each represented in the numerator as three Planck-length-long segments with 2.17 x 10⁻⁸ kg of mass, moving at the speed of light. The value 1.60 x 10⁻¹⁹ Coulombs (C) represents the fundamental unit of electric charge, positive or negative. That number squared stands for two interacting electric charges. The value 6 x 137c represents the contribution of the flow of time to the interaction between the charges—see the three curved lines representing the flow of t^+ energy and the three curved lines representing the flow of t^- energy.*

est it can be—the Planck length—carrying the greatest amount of mass it can—the Planck mass—and moving at the speed of light (c). In electric attraction, the two primary lines move toward each other; in electric repulsion, they move away from each other.

In the denominator, the value 1.60×10^{-19} Coulombs represents the fundamental unit of electric charge, positive or negative. That number squared stands for two interacting electric charges—they may be undergoing electric attraction or repulsion. The value 137c (137 times the speed of light) represents the contribution of the flow of time to the interaction between the charges. (The number 137 comes from a known physical quantity [the fine structure constant].) That flow is related to the exchange of t^+ and t^- between the opposing secondary field lines of the interacting charges, with the temporal particles moving 137 times faster than light. In electric attraction, this flow helps to further unite the merging field lines, serving as a coupling agent. It is multiplied by six because each of the three segments in one secondary line can be considered to pass temporal energy to its counterpart in the other secondary line. Thus, there are three flows of t^+ and three flows of t^- (see *Figure 10.12*). Interestingly, variations of this coupling are present in the strong, weak, and nuclear contact forces, as well as the universe as a whole, which of course includes gravity (*Figure 10.13*). The various versions have differing strengths. Note that the forces can occur no faster than the speed of light, but the flow of time adds to their strength—the field lines are coupled through the flow of time between them.

Although in electric attraction, the flow of time helps bring the initial primary lines together, in electric repulsion, it helps drive them apart. The middle section of *Figure 10.14* shows the contribution of the flow of time to electric repulsion between two negative charges. The yellow components of the electrons' field lines always face each other. As with other yellow elements, they produce and repel t^+. However, because both interacting lines are yellow, neither will accept this t^+ the way a blue element would. As a result, a wall of t^+ builds up between them, and the field lines, in a sense, push against this wall, away from each other. This is similar to two people in a pool lying on little plastic rafts bringing their feet together and pushing themselves apart. The same

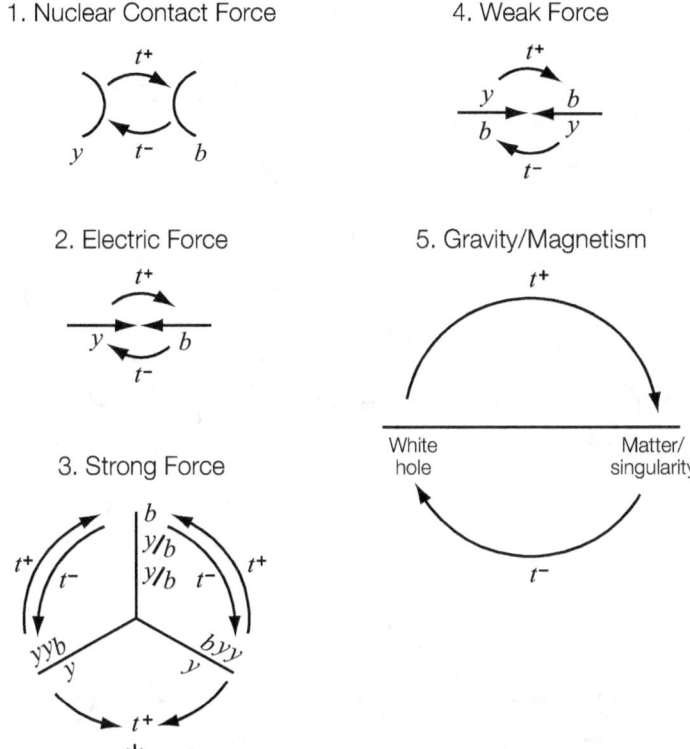

Figure 10.13 *Variations of the coupling agent in the electric force are present in the strong, weak, and nuclear contact forces, as well as gravity/magnetism. Note that the two arrows in each example should be thought of as existing simultaneously, rather than one occurring then the other. In the example shown for the strong force, the focus is on the flow to time between the electrical elements only; also, there is a repulsive "coupling" (marked by an asterisk), which is described in Figure 10.14.*

would occur for two blues lines facing each other, except that they would be pushing against a wall of t^-. As noted above for the other forces, the force of repulsion cannot occur faster than the speed of light; the flow of time adds to the strength of the force rather than its speed.

The speed of light squared (c^2) is the conversion of t^+ to t^0 and t^0 to t^- (or the reverse) by the field lines. Basically, from the standpoint of TET, K represents the reduction of a common field line between two electrically charged particles (represented by the expression in the denominator of the equation in *Figure 10.12*) into an interaction between the initial primary lines emanating from the electrically charged particles (represented by the expression in the

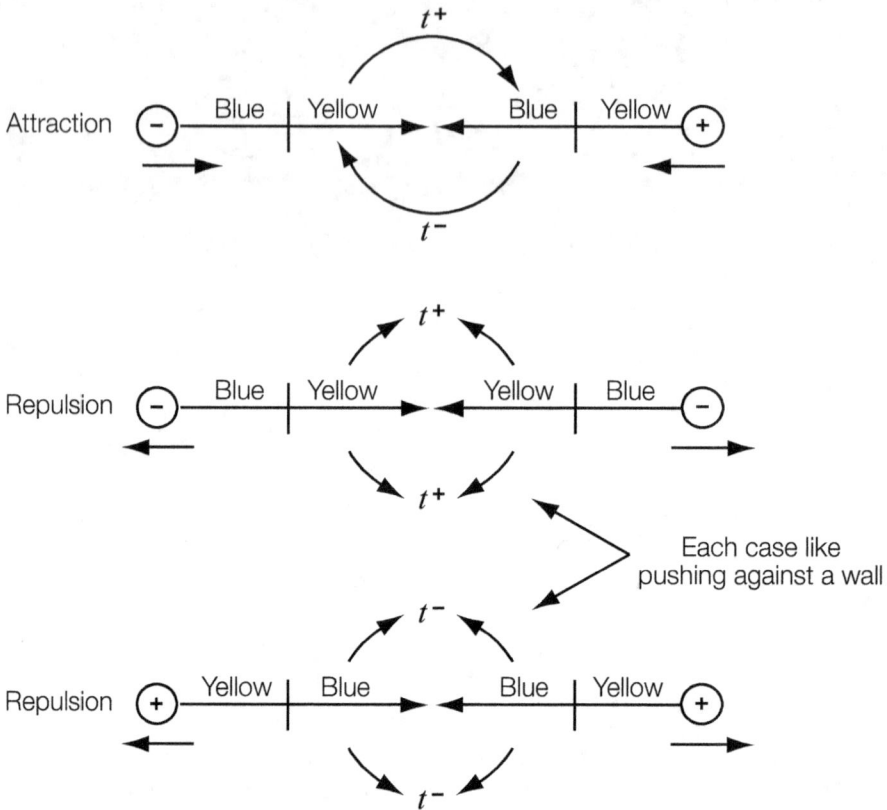

Figure 10.14 *Although in electric attraction, the flow of time helps bring the initial primary lines together, in electric repulsion, it helps drive them apart. The interacting field lines of two negatively charged particles push against a wall of t^+, whereas the interacting field lines of two positively charged particles push against a wall of t^-.*

numerator of the equation). The primary lines may be moving toward each other (in electric attraction) or away from each other (in electric repulsion), at the speed of light. Although the examples above focus on the Planck level, including the charged particles themselves, the value of K is the same in electrical interactions between charges at higher scales, as well.

Note that in Coulomb's Law, electric charges are regarded as dimensionless, point particles. From the perspective of TET, these point particles are elementary particles in their most compressed state, which is what they would be when moving at or, more technically, very close to the speed of light. In the idealized model of *Figure 10.12*, the particles are accelerated to near the speed of light by the electric force, in which case they would have to have started in a highly expanded, low-mass state, as high mass would inhibit such acceleration. Thus, in TET, the particles are not dimensionless, although they are indeed small. For an electron, it is the state in which its diameter is the Planck length and its frame consequently consists of three Planck-length-long polarized segments, with each initial primary line emanating from this particle also consisting of three Planck-length-sized polarized segments—although in the idealized model, each subsequent line also contains just three segments of this size. As noted earlier, K, in TET, represents the engine of the electric force—the phenomena that create the action. The greater equation of Coulomb's Law deals with the magnitude of the force.

The Vacuum Energy Problem

It is has been understood for some time that otherwise empty space is filled with energy—referred to as the vacuum energy or the cosmological constant. This energy causes the universe to expand. In General Relativity (GR), it is represented by a positive energy term, associated with a negative pressure, that drives the expansion.[18] It has a measured value of approximately 10^{-27} kg/meter^3 [19,20] but a theoretical value of approximately 10^{96} kg/meter^3, which is about 10^{123} times larger than the measured value.

In TET, the vacuum energy is temporal energy, specifically the temporal energy bounced back and forth between interior space and exterior space and the blue "half" of interior space and its yellow "half." It is dark energy and indeed plays a role in causing the universe to expand/accelerate as explained in chapter 9. Its association with a positive energy term in GR is likely a reflection of exterior space's reaction to it when it is in the form of t^+, as exterior space expands due to its reaction with t^+. Thus, t^+ energy is likely the "positive energy" driving the expansion in GR. This positive energy is associated with a negative pressure in our space because exterior space pulls on interior space as it expands, preventing interior space from collapsing. Using a sealed, deflated balloon for comparison, the balloon would experience positive pressure if it were expanded by being injected with air; it would experience negative pressure if it were expanded by something pulling on its skin from the outside, which in a sense, is what exterior space does to interior space—hence the negative pressure associated with interior space's expansion.

Note that the filling up of a space with temporal particles does not in itself cause the space to expand. The temporal particles simply stabilize areas where true-vacuum regions develop within that space. If exterior space were not repelled by t^+ and did not, as a consequence, pull on interior space, both spaces would collapse, absorbing any temporal particles within them along the way.

One way the theoretical value of the energy (at about 10^{96} kg/meter^3) is attained is by assuming a volume of otherwise empty space with a length, width, and height of 1 Planck length each, is filled with the energy of the Planck mass (2.17×10^{-8} kg)—that is 2.17×10^{-8} kg divided by the Planck length cubed. From the TET perspective, the Planck mass represents the highest amount of mass a fully polarized line of interior space that is 1 Planck length long can have. However, there is no reason to assume that every Planck-scale volume of otherwise empty space has a fully polarized line in it, and thus, there is no reason to assume that every volume of space this size, being otherwise empty, has the Planck mass inside of it. In TET, only one temporal particle is assumed to exist in such a volume. Using the standard density equation (density equals mass divided by volume) and setting the density equal to the measured value of 10^{-27} kg/meter^3 leads to a Planck-scale volume of space containing approximately 10^{-131} kg. Considering it would be the mass equivalent of a single temporal particle, 10^{-131} kg is not an unusual number. A temporal particle is extremely small, and so is this mass.

Interestingly, there is a way of regarding the vacuum energy as being zero in TET, as well. Some of the t^+ composing this energy is constantly bounced back and forth between interior space and exterior space, and some is constantly bounced back and forth between the blue "half" of interior space and its yellow "half." Thus, the total of the t^+ energy spends half its life as t^+ and half as t^-. As discussed earlier regarding renormalization, these two states can, at least mathematically, be thought of as canceling each other, leading to a vacuum energy value of exactly zero.

11

PARTICLE BEHAVIORS

This chapter examines three important behaviors exhibited specifically by QM particles. They are their ability to behave like particles and waves, tunnel through barriers, and seemingly alter each other's behavior even when separated by great distances.

Particle/Wave Duality

The particle/wave duality of QM particles comes through in an experiment known as the double-slit experiment. In this experiment, light is beamed through two tiny slits in a screen toward a back screen. The result is the development of an interference pattern on the back screen, which appears as a continuous band of alternating bright and dark areas along the length of the screen. This experiment proved that light behaves as waves, with bright areas being places where the waves interfered constructively, and dark areas being places where the waves interfered destructively. However, later experiments, such as the photoelectric effect, seemed to contradict the idea of light as waves. In such experiments, light behaves as if it were made of particles—as if it were little pellets that could strike other objects, like two balls hitting one another. A beam of pellets shot through two slits in a wall would not create an interference pattern. Instead, they would create two spots on the back wall corresponding to the slits the pellets came through.

Matter particles, such as electrons, also seem to have this dual particle/wave nature. At times, they behave very much like particles, altering the position and momentum of other particles they strike, yet a beam of electrons shot through a double-slit screen will create an interference pattern, rather than two individual spots on the back screen corresponding to the slit they came through. Mysteriously, even when photons and electrons are painstakingly shot one at a time through the slits, an interference pattern still develops—despite the lack of another particle to interfere with it. Yet equally mysterious is the fact that if a detector is placed at one of the slits to show which slit that lone particle traveled through, the interference pattern disappears. It is as if knowing which slit the particle traveled through inhibits its ability to create the interference pattern.

In TET, the dual particle/wave nature of photons, electrons, and similar particles comes from the nature of their frames and the energy those frames carry. With regard to photons, their frames are indeed always in the form of waves, but those waves also behave like little packets of energy—that is, particles—through the energy within the frames (blue/positive and yellow/negative energy) and the collective energy of the temporal particles associated with those frames. With regard to electrons, they are indeed particles, but their frames undulate like waves when they move (see *Figure 3.30*). Thus, photons and electrons can behave as particles or waves depending on the circumstances they are in.

Interestingly, the interference pattern produced on a screen through the firing of such particles as photons and electrons is produced with the help of space and time from the standpoint of TET. Temporal particles are attracted to energy, moving in unison with the source of that energy. Thus, as the waves of photons and electrons undulate as they propagate through space, the temporal particles responding to this energy will also develop an undulating/wave pattern. When temporal particles converge on an area of space, they compress it, as if they were molding it. Therefore, as waves form in the field of temporal particles in the vicinity of the photons and electrons, space itself takes on a wave pattern in this area. And as is the case with any waves that are spread out and that pass through two slits within a boundary, these waves of space and time develop an interference pattern on the back side of the two slits.

When a photon or electron passes through one of the slits, its trajectory is altered by the space/time interference pattern behind the slits, hitting the back screen according to this pattern. The pattern quickly dissipates due to the motion of the particle on the other side of the double-slit screen—that is, a new wave pattern forms—but by that point, its motion has already been affected. Its course is affected just past the threshold of the slit it came through. When multiple particles are sent through at once, they develop an interference pattern on the back screen quickly—again, with the particles hitting the back screen according to the interference pattern of the space/time surrounding them. When particles are sent through the slits one by one, they still eventually distribute themselves on the back screen according to an interference pattern, because it takes only one traveling photon or electron to cause space and time to develop a wave nature/interference pattern in a double-slit experiment.

The placement of a detector at one of the slits adds energy to that area. Because temporal particles respond to energy, many converge on the detector, which breaks up or prevents a space/time wave pattern from developing in that area. With no waves produced at one of the slits, no interference pattern develops behind the slits. When photons or electrons are sent through the slits, they hit the back screen in a similar way that little pellets would, with just two spots forming on the back screen where the particles landed.

Quantum Tunneling

In quantum tunneling, particles are able to traverse seemingly impenetrable barriers. In Quantum Mechanics (QM), this is attributed to the uncertainty in a particle's position. From this standpoint, there is a small, but definite chance for the particle to be found on the other side of the barrier, even though it technically does not have energy to cross the barrier. In TET, quantum tunneling comes about through spatial compression.

For example, consider a proton and neutron to interact through the nuclear contact force, one of the strongest forces known to exist. In TET, the contact force comes about through a temporally symbiotic relationship between a yellow segment in the frame of one nucleon and a blue segment in the frame of another. The nucleons within a light nucleus are about the same size; that is, they have about the same amount of compression due to the surrounding temporal particles. As a result the contact force between them is particularly strong; it would be extremely difficult to separate them. In a heavy nucleus, however, the areas deep inside it have more compression than the areas toward its outer edges. Nucleons in the more compressed areas will themselves be more compressed. The differences in the sizes of the nucleons in the nucleus allow some of them to tunnel out of the nucleus—that is, to leave it. An example of this is called alpha-decay, in which an alpha particle (consisting of two neutrons and two protons) leaves the nucleus it was once a part of.

Figure 11.1 illustrates, in an exaggerated way for effect, an interaction between two nucleons of equal compression and an interaction between a nucleon in a more compressed area of a nucleus and one in a less compressed area of the nucleus. In the latter case, the contact force still exists between the nucleons, but the blue segment, in this example, is in a highly expanded

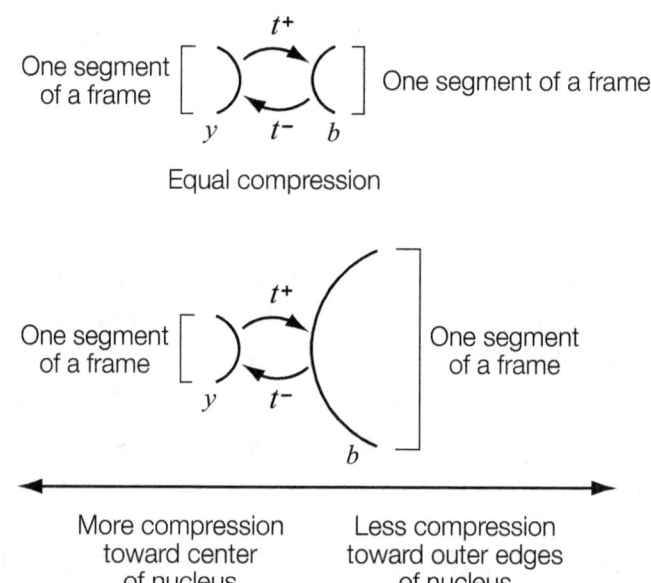

Figure 11.1 *When two nucleons have equal or near-equal compression, the contact force between them is exceptionally strong. When one is less compressed, the contact force does not have as strong a hold on that particle, allowing it to tunnel out of the nucleus.*

state compared to the yellow segment. Because of this, the contact force between the nucleons is weakened, which allows the more expanded particle to break away, taking perhaps a few other nucleons with it.

Spatial compression and thus tunneling can also occur when particles move fast or are highly excited. For example, under certain conditions, an electron in a metal, if excited, can be made to leave the metal, jump across space, and land in another material. In TET, temporal particles converge on the electron due to its extra energy, compressing the particle. When charged particles are compressed, the number of field lines emanating from them is reduced (see *Figure 6.1* and legend). With a reduction in the field lines, there is a weakening of the interactions the electron is engaged in, giving it more freedom to break away. That is, the reduction in field lines leads to a weakening of the attractive forces holding onto the electron.

Quantum Entanglement

Two QM particles are said to be entangled when, after interacting and then separating, a measurement of one particle's spin will immediately determine the spin of the other particle, regardless of the distance between them. If the spin of one particle is determined to be moving clockwise, the spin of the other will somehow mysteriously, and instantly, "know" it should be moving counterclockwise. In QM, it is considered to be equally likely for the first spin to have been measured to be clockwise as counterclockwise. Inherent in this idea is the QM concept of spin not existing in any definite state until it is measured.

In TET, however, a particle has a definite spin even before it is measured; however, if its spin moves in a circle, the clockwise or counterclockwise nature of that spin will depend on the particle's direction of travel and the observer's orientation. *Figure 11.2* shows TET's description of two electrons, called a Cooper pair, which have been known to undergo entanglement. In this example, both originally spin counterclockwise. Following their interaction, the electrons move in opposite directions. In the figure, electron A is moving into the page—designated as A-prime, with an "x" in the middle indicating its

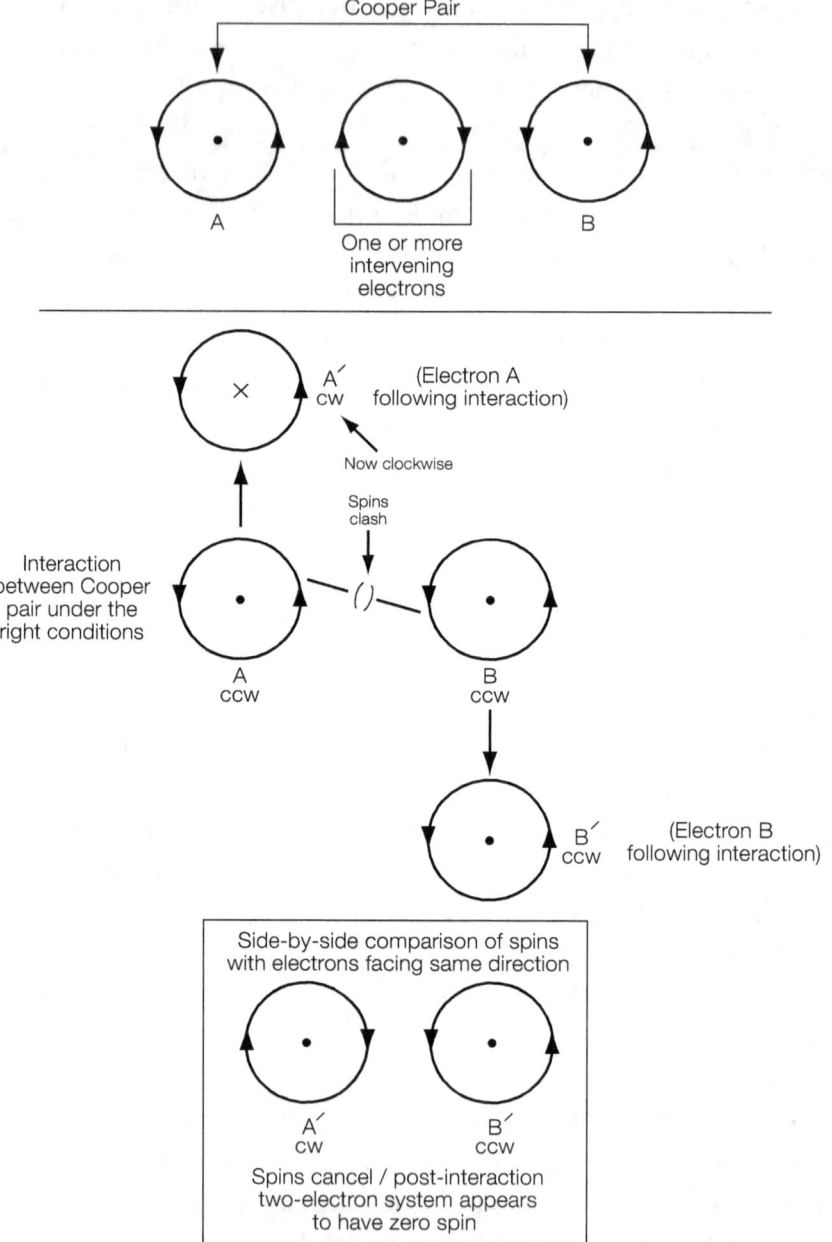

Figure 11.2 *If one particle in an entangled pair, such as a Cooper pair, is measured to have a clockwise spin, the other will be measured to have a counterclockwise spin. In TET, the spins of particles preexist; they do not appear only when measured. However, the direction of spin (clockwise or counterclockwise) will depend on a particle's direction of travel (see Spin section in chapter 3).*

motion into the page. Electron B is moving toward the reader—designated as B-prime, with a dot in the middle. (The "x" is like looking at the back of an arrow as it travels away from you, and the dot is like looking at the tip of an arrow as it travels toward you.)

Electron A, now as electron A-prime, has a clockwise spin, because spin must be reckoned from the direction of motion; with this electron moving into the page, you would have to imagine yourself standing between the electron and the page, watching the particle as it moved toward you. Electron B, now as electron B-prime, continues to have a counterclockwise spin. The situation could exist in reverse, with electron A moving toward the reader and spinning counterclockwise following the interaction and B moving toward the page and spinning clockwise. Also of course, the original spins of the particles could have been clockwise instead of counterclockwise. Regardless of these different possibilities, the electrons would be measured to have opposite spins follow-ing their interaction. But again in TET, those spins preexist. In contrast to QM, there is no need in TET for one of the electrons to somehow "tell" its counterpart, some distance away, what spin it has once it has been measured, so that its partner can instantaneously adopt the opposite spin. This is also the case for other particles, including atoms.

Note that, taken together, the end-products—electron A-prime and electron B-prime—have zero spin between them. This is because one spin is moving clockwise and the other counterclockwise (see box in *Figure 11.2* and paired electrons in *Figure 7.5*). In a certain way, this information suggests that the pre-interaction state of the electrons was also spin zero—that is, extrapolating backward, if the end-products together are spin zero, their original state would seem to have been spin zero. In TET, however, the original state of the electrons involved them having spins that clashed not canceled, and it is this clashing that leads to the interaction between them.

SUMMARY

Temporal Energy Theory is a theory of time, but it can also be regarded as a theory of space, time, and energy—the three factors from which everything in the universe is made. Temporal Energy Theory is also an alternative Grand Unification Theory, a Theory of Quantum Gravity (with temporal particles being the quanta of gravity), and a Theory of Everything, in that it unites General Relativity and Quantum Mechanics. It indicates that the electric force is at the heart of gravity, magnetism, and the weak and strong nuclear forces, as well as the nuclear contact force.

An important process in the theory is the convergence of temporal particles on sources of energy. However, these particles may not be responding to energy per se, but more specifically spatial fluctuations, with them responding even to the spatial fluctuations that they themselves cause. Although, even if this is the case, saying that they converge on energy is more expedient. Also, Temporal Energy Theory has much to do with fields—electric, weak, and strong—and thus is a new field theory. The ultimate field in the theory, of course, is the field of temporal particles filling interior space. The field lines linking temporal particles to one another and also to particle frames and whole vacuums represent yet another type of field in the theory.

Temporal Energy Theory involves several important forces in addition to the electric, weak, strong, gravitational, magnetic, and nuclear contact forces. Six of these forces can be considered its fundamental forces, although the term is used loosely (*Table 12.1*). The spatiotemporal force allows space to become temporally polarized—particularly to generate electric, weak, and strong fields. It is also likely the force at the heart of the Big Bang and tachyon/temporal particle creation when true-vacuum regions develop in space. The next force is the temporal-conversion force, allowing space to convert one type of temporal particle into another to maintain its own polarization. After this, comes

Table 12.1 Summary of TET Forces

Spatiotemporal Force	Electric Force
Temporal Conversion Force	Weak Force
Temporal Particle Force	Strong Force

the temporal-particle force, which binds temporal particles together and also binds these particles to particle frames or whole vacuums. The attraction and repulsion of temporal particles by space also likely occur through this force. The final three are the electric, weak, and strong forces. Gravity, magnetism, and the nuclear contact force are due to combinations of the spatiotemporal, temporal-conversion, temporal-particle, and electric forces—with the electric force being at the center of the activity, in that the spatiotemporal force creates an electric field line, which then binds with t^+ through the temporal-particle force and converts it to t^- through the temporal-conversion force, allowing gravity, magnetism, and the nuclear contact force to occur.

CONCLUSION

The greatest source of energy is time. Arguably, in the early universe the answer would have been space, as it is likely the origin of time, but in the here and now, it is time. As a physical object, time is an energy field that spans the cosmos. It surrounds us, penetrates us, and binds everything and everyone together. Time gives the universe substance—it gives mass to matter. Time is the solid ground under our feet, the clothes on our backs, the food we eat, our skin and bones, and the sunlight that shines on us, to name just a few examples. As a process, time sustains us. It is the breath of life for every rock, tree, and creature. Through time, everything lives, and everything is connected.

It is interesting that time has seemed so mysterious and intangible to humanity for so long, and yet if Temporal Energy Theory is correct, it is actually very conspicuous and palpable. It is further interesting that, not only may time be at the root of the physical world, it may be at the root of the mental world, as well. An important concept promoted by René Descartes is, "I think; therefore, I am." This is a reasonable idea, and it is equally reasonable that the content of our very first thought is about time—a raw understanding that this moment is somehow different from the previous one. It is likely what we know first and what therefore provides us with self-awareness. The link to the mind is clear— the basic time of electrons and quarks leads to thermodynamic time, which in turn leads to psychological time.

Temporal Energy Theory, if proven correct, cannot be regarded as a theory made from scratch; rather, it is an extension of past ideas dealing with the forces of nature. Isaac Newton provided many profound and lasting thoughts on gravity. Albert Einstein followed up on Newton's thoughts, providing a more precise description of gravity in General Relativity, and he helped lay the foundation of Quantum Mechanics, along with a long list of other distinguished scientists, for describing electromagnetism and the nuclear forces. Temporal Energy

Theory is an endeavor to combine General Relativity and Quantum Mechanics and thus the forces they describe, picking up from points where the other theories leave off.

Hopefully, Temporal Energy Theory succeeds in combining General Relativity and Quantum Mechanics. If this is proven to be the case, the theory will have redeemed Newton's idea of a universal background clock, without taking anything away from Einstein's idea that time is relative. From the standpoint of Temporal Energy Theory, they are both right. Hopefully, the theory also succeeds in shedding additional light on many other scientific concepts and questions and, in general, helps to further advance human thought and understanding about the universe.

Is Temporal Energy Theory an accurate theory of time? Does it successfully combine General Relativity and Quantum Mechanics? Is it the Theory of Everything?

Time will tell.

REFERENCES AND
SUGGESTED READINGS

References

1. Wheeler JA. *A Journey into Gravity and Spacetime.* Scientific American Library, New York, 1990, pp. 11–12.

2. Einstein A. *The Meaning of Relativity.* 5th Ed. Princeton University Press, Princeton, 1922, 2005, pp. 64 and 81.

3. Wootters WK. Why Things Fall. *Foundations of Physics.* Vol. 33, No. 10, 2003, pp. 1549–57.

4. Misner CW, Thorne KS, Wheeler JA. *Gravitation.* Macmillan, New York, 1973, p. 32.

5. D'Abro A. *The Evolution of Scientific Thought from Newton to Einstein.* Boni and Liveright, 1927, p. 301.

6. Jagerman L. *The Mathematics of Relativity for the Rest of Us.* Trafford Publishing, 2001, p. 224.

7. Nissani N, Leibowitz E. Global Energy-Momentum Conservation in General Relativity. *International Journal of Theoretical Physics.* Vol. 28, No. 2, 1989, p 235–245.

8. Lynden-Bell D, Katz J. Energy (Non-) Conservation near Black and White Holes. *Classical Quantum Gravity.* Vol. 8, 1991, pp. 403–6.

9. Halliday D, Resnick R, Walker J. *Fundamentals of Physics.* 4th Ed. John Wiley & Sons, Inc., New York, 1993, p. 147.

10. See reference 9, p. 655.

11. Rohrlich F. Time in Classical Electrodynamics. *American Journal of Physics.* Vol. 74, No. 4, 2006, pp. 313–5.

12. Lawrence Berkeley National Laboratory—A U.S. Department of Energy National Laboratory Operated by the University of California. Charge, Parity, and Time Reversal (CPT) Symmetry. Available at: http://www.lbl.gov/abc/wallchart/chapters/05/2.html.

13. Rees M, Natarajan P. A Field Guide to the Invisible Universe. *Discover.* December 2003, pp. 42–49.

14. Froggatt CD, Nielsen HB. *Origin of Symmetries.* World Scientific, River Edge, NJ, 1991, p 40.

15. Thorne K (Kruglinski S, interviewer). The Discover Interview—Kip Thorne. *Discover.* November 2007, pp 51–53.

16. Hawking S. *A Brief History of Time.* Bantam Books, New York, 1990, pp. 99–113.

17. National Institute of Standards and Technology (NIST). The NIST Reference on Constants, Units, and Uncertainty. Available: http://physics.nist.gov/cuu/index.html.

18. Straumann N. On the Cosmological Constant Problems and the Astronomical Evidence for a Homogeneous Energy Density with Negative Pressure. Feb. 5, 2008. Available: http://arxiv.org/abs/astro-ph/0203330.

19. Longair MS. *Astronomy and Astrophysics Library: Galaxy Formation.* 2nd Ed. Springer, New York, 2008, pp 623.

20. Carroll SM, Press WH, Turner EL. The Cosmological Constant. *Annual Review of Astronomy and Astrophysics.* Vol. 30, 1992, pp. 499–542.

Suggested Readings

DeWitt BS. Quantum Theory of Gravity I. The Canonical Theory. *Physical Review.* Vol. 160, 1967, pp. 1113–48.

Feinberg J. Self-Adjoint Wheeler-DeWitt Operators, the Problem of Time and the Wave Function of the Universe. Mar. 31, 1995. Available: http://arxiv.org/abs/hep-th/9503073.

Garcia-Bellido J, Linde A, Wands D. Density Perturbations and Black Hole Formation in Hybrid Inflation. May 15, 1996. Available: http://arxiv.org/abs/astro-ph/9605094.

Montesinos M. The Double Role of Einstein's Equations: As Equations of Motion and as Vanishing Energy-Momentum Tensor. Feb. 7, 2008. Available: http://arxiv.org/abs/gr-qc/0311001.

Particle Data Group. *Review of Particle Physics*. 2010 Ed. Available: http://pdg.lbl.gov.

INDEX

(Note: "definition" indicates first mention of a term used throughout book.)

www.ingramcontent.com/pod-product-compliance
Lightning Source LLC
Chambersburg PA
CBHW071255220526
45468CB00001B/141